W9-ADM-114

DISPOSSESSING *the* WILDERNESS

DISPOSSESSING *the* WILDERNESS

Indian Removal and the Making of the National Parks

Mark David Spence

New York Oxford
Oxford University Press
1999

Oxford University Press

Oxford New York
Athens Auckland Bangkok Bogotá Buenos Aires Calcutta
Cape Town Chennai Dar es Salaam Dehli Florence Hong Kong Istanbul
Karachi Kuala Lumpur Madrid Melbourne Mexico City Mumbai
Nairobi Paris São Paulo Singapore Taipei Toyko Toronto Warsaw

and associated companies in
Berlin Ibadan

Published by Oxford University Press, Inc.
198 Madison Avenue, New York, New York 10016

Oxford is a registered trademark of Oxford University Press

Library of Congress Cataloging-in-Publication Data
Spence, Mark David.
Dispossessing the wilderness : Indian removal and the making of the
national parks / by Mark David Spence.
p. cm.
Includes bibliographical references.
ISBN 0-19-511882-0
1. Indians of North America—Relocation—West (U.S.)
2. Wilderness areas—Government policy—United States. 3. National
parks and reserves—Government policy—United States. 4. Nature
conservation—Social aspects—United States. I. Title.
E98.R4S64 1999
978'.00497—dc21 98-27456

1 3 5 7 9 8 6 4 2

Printed in the United States of America
on acid-free paper

Wallace Spence, may this help to close the circle.

Therese Mary Johnson, for making this possible.

Amanda Kate Allaback, for seeing it to the end.

ACKNOWLEDGMENTS

MANY PEOPLE HAVE ENCOURAGED and supported my long fascination with the subjects of this book. My family and friends have all contributed more than I can possibly acknowledge here. Some expressed genuine interest in my work while others just politely feigned a mild concern, but all have tolerated with much patience and good humor the social deficiencies of a young academic. My very deepest thanks to all.

Colleagues and mentors have also become friends along the way, and I would like to acknowledge a number of professional and personal debts that I shall long struggle to repay. Michael O'Connell of the University of California, Santa Barbara, first introduced me to the challenges and rewards of serious historical study. I owe much to the University of California, Los Angeles, where Ellen DuBois, Susanna Hecht, Melissa Meyer, and Theodore Porter provided a great deal of encouragement and insight. Tanis Thorne also gave a portion of the manuscript a valuable critique at an important juncture. Likewise, Hal Rothman, Elliott West, Susan Rhoads Neel, Stephen Aron, and Peter Nabokov offered kind criticism on various portions of the manuscript. Jim Sherow read an entire draft, as did a number of anonymous referees, and all provided invaluable suggestions for improving this book. Besides offering wonderful advice and commentary, Marguerite Shaffer, Louis Warren, and Karl Jacoby generously shared their own scholarship. My greatest thanks go to Norris Hundley Jr. of the University of California, Los Angeles. His diligence, wise advice, and elegant use of language make him a model advisor. I can only hope my work justly reflects his great attention and care.

A number of institutions supported the research and writing of this work, including the Andrew W. Mellon Foundation, the Huntington Library, the UCLA Department of History, the William Andrews Clark Memorial Library, the UCLA American Indian Studies Center, and the Carey McWilliams Awards Fund. All illustrations were acquired with the financial assistance of Knox College.

Portions of the material in chapters 5 and 6 appeared in *Environmental History* (1996), copublished by the American Society for Environmental History and the Forest History Society in association with Duke University. Likewise, a previous manifestation of the material in chapters 7 and 8 appeared in the *Pacific Historical Review* (1996), published by the University of California Press.

I have had the good fortune of conducting research in a number of excellent archives and libraries in some extraordinary places. Beth Gunnison and Deirdre Shaw at Glacier National Park, Tom Tankersley and Lee Whittlesey at Yellowstone National Park, and Linda Eade at Yosemite National Park graciously shared their knowledge of park history and pointed out dozens of crucial documents that I could never have found on my own. I also benefited greatly from the expertise of Peter Blodgett at the Huntington Library, Caroll Sommer at the William Andrews Clark Memorial Library, Magdalen Medicine Horse at the Little Big Horn College Archives, and Joyce Justice at the Denver Federal Archives and Records Center. Likewise, I owe a great debt to the expert staffs of the National Archives I and II in Washington, D.C., and College Park, Maryland, the Seattle Federal Archives and Records Center, the Bancroft Library, the Merril G. Burlingame Special Collections at Montana State University, the Montana Historical Society, the Southwest Museum in Los Angeles, and the Special Collections of the University of California, Los Angeles.

Some of my deepest thanks go to Barbara Sutteer, Ted Hall, David Ruppert, and Craig Bates of the National Park Service; Chief Earl Old Person, Mike Swims Under, Marvin Weatherwax, Curly Bear Wagner, Joyce Spoonhunter, Ramona Hall, Darrell Kipp, and other members of the Blackfeet Nation; Genevieve Edmo, Tony Galloway, John Fred, Larry Bagley, Diane Yupe, and other members of the Shoshone-Bannock Tribes; Jay Johnson and other members of the American Indian Council of Mariposa County; and Mick Old Coyote, Darrel Old Horn, Francis and Celise Stewart, John Pretty On Top, and other members of the Crow Tribe. All have generously provided guidance, assistance, and invaluable knowledge.

Finally, I would like to dedicate this book to three people: Wallace Spence, whose strange life long ago cemented my own connection to many of the places and personalities discussed in this study; my mother, Therese Mary Johnson, whose integrity, love, and pride are the source of endless inspiration; and my wife, Amanda Kate Allaback, whose love and encouragement made this work just one of life's many great joys.

CONTENTS

INTRODUCTION
From Common Ground 3

1
Looking Backward and Westward:
The "Indian Wilderness" in the Antebellum Era 9

2
The Wild West, or Toward Separate Islands 25

3
Before the Wilderness:
Native Peoples and Yellowstone 41

4
First Wilderness:
America's Wonderland and Indian Removal
from Yellowstone National Park 55

5
Backbone of the World:
The Blackfeet and the Glacier National Park Area 71

6
Crowning the Continent:
The American Wilderness Ideal and Blackfeet
Exclusion from Glacier National Park 83

7
The Heart of the Sierras, 1864–1916 101

8
Yosemite Indians and
the National Park Ideal, 1916–1969 115

CONCLUSION
Exceptions and the Rule 133

Notes 141

Index 181

DISPOSSESSING *the* WILDERNESS

INTRODUCTION

From Common Ground

> I wonder if the ground has anything to say? I wonder if the ground is listening to what is said? I wonder if the ground would come alive and what is on it?
>
> We-ah Te-na-tee-ma-ny,
> or "Little Chief" (Cayuse), 1855[1]

SHORTLY AFTER THE ESTABLISHMENT OF Badlands National Monument in 1929, the Oglala Sioux spiritual leader Black Elk expressed profound consternation with the idea of wilderness preservation. For him, the creation of the national monument adjacent to his home on the Pine Ridge Reservation in South Dakota seemed only to confirm a disturbing trend. Wind Cave National Park had already been established in the nearby Black Hills, and large areas of land surrounding the park had recently been incorporated into a national forest. Remembering his youth and the time he spent in these areas, Black Elk recalled that his people "were happy in [their] own country, and were seldom hungry, for then the two-leggeds and the four-leggeds lived together like relatives, and there was plenty for them and for us." Although a considerable portion of this Sioux country received federal protection, native peoples were largely excluded from their former lands. As Black Elk observed, the Americans had "made little islands for us and other little islands for the four-leggeds," and every year the two were moving farther and farther apart.[2] In short, Black Elk understood all too well that wilderness preservation went hand in hand with native dispossession.

The dual "island" system of nature preserves and Indian reservations did not originate in the 1920s. At least until Black Elk's early childhood, Americans generally conceived of the West as a vast "Indian wilderness," and they rarely made a distinction between native peoples and the lands they inhabited. Consequently, the earliest national park advocates hoped to protect "wild" landscapes and the people who called these places home. Preservationist efforts did not succeed until the latter half of the nineteenth century, however, when outdoor enthusiasts viewed wilderness as an uninhabited Eden that should be set aside for the benefit and pleasure of vacationing Americans. The fact that Indians continued to hunt and light purposeful fires in such places seemed only to demonstrate a marked inability to appreciate natural beauty. To guard against these "violations," the establishment of the first national parks necessarily entailed the exclusion or removal of native peoples.

The transition in American conceptions of wilderness resulted from several deeper trends in U.S. society and politics. The powerful sense of national destiny that accompanied both the Mexican War and the Civil War, the increased activism of the federal government during Reconstruction, the growth of western tourism, and the widespread sentimentalism for a "vanishing" frontier profoundly shaped the ways that Americans would perceive the "New West" for several decades. For many people, the processes of conquest and nation building seemed to alter the essential nature of the region; through a sort of patriotic transubstantiation, a number of western landscapes quickly became American Canterburys. More than great "pleasureground[s] for the benefit and enjoyment of the people," the first national parks were places where summer pilgrims could go to share their national identity and an appreciation for natural beauty.[3] Much as they still do today, Yosemite Valley, the Grand Canyon of the Yellowstone, and the ragged peaks of the northern Rocky Mountains provided the basic elements of a scenic anthem that praised the grandeur and power of the United States.

The idealization of uninhabited landscapes and the establishment of the first national parks also reflect important developments in late-nineteenth-century Indian policy. Much as the conquest of the West reshaped ideas about wilderness, it also led to the creation of an extensive reservation system. Ultimately, these isolated patches of land came to represent the final refuge of the American Indian, and by the late 1860s and early 1870s, Americans regarded reservations, rather than the "wilderness," as the appropriate place for all Indians to live. These sentiments changed somewhat in the following decades, when self-described friends of the Indian sought to dismantle the reservations and assimilate native peoples into American society. While such "friends" argued that an Indian's place was not on the reservation, they asserted even more emphatically that an Indian's place was not in the wilderness—except on the odd chance that one had become a "civilized" tourist.

Changing ideas and policies did not make native peoples disappear, however, nor did they make wilderness uninhabited. Although the creation of the first national parks coincided with efforts to restrict Indians to reservations and assimilate them into American society, native use and occupancy of park lands often continued unabated. A basic argument of this book is that uninhabited wilderness had to be created before it could be preserved, and this type of landscape became reified in the first national parks. In particular, I focus on the policies of

Indian removal developed at Yosemite, Yellowstone, and Glacier national parks from the 1870s to the 1930s. These parks are especially relevant for three reasons: first, each supported a native population at the time of its establishment; second, the removal of Indians from these parks became precedents for the exclusion of native peoples from other holdings within the national park system; and third, as the grand symbols of American wilderness, the uninhabited landscapes preserved in these parks have served as models for preservationist efforts, and native dispossession, the world over.[4]

Generations of preservationists, government officials, and park visitors have accepted and defended the uninhabited wilderness preserved in national parks as remnants of a priori Nature (with a very capital *N*). Such a conception of wilderness forgets that native peoples shaped these environments for millennia, and thus parks like Yellowstone, Yosemite, and Glacier are more representative of old fantasies about a continent awaiting "discovery" than actual conditions at the time of Columbus's voyage or Lewis and Clark's adventure.[5] For the most part, these romantic visions of primordial North America have contributed to a sort of widespread cultural myopia that allows late-twentieth-century Americans to ignore the fact that national parks enshrine recently dispossessed landscapes.

In the past few years, a number of scholars have argued that wilderness is not an absolute condition of Nature but is instead a fairly recent American invention.[6] While I share the conviction that wilderness is both a historical and cultural construct, I believe that such a definition requires an examination of the events and processes that led to the creation of this particular artifact. Doing so should also make plain the manner in which popular conceptions of certain wilderness areas have precluded alternate visions of the same landscapes. Ultimately, an understanding of the context and motives that led to the idealization of uninhabited wilderness not only helps to explain what national parks actually preserve but also reveals the degree to which older cultural values continue to shape current environmentalist and preservationist thinking.

The American wilderness ideal, as it has developed over the last century, necessarily includes a number of strange notions about native peoples and national parks. In the rare instances that park literature even mentions Indians, they tend to assume the unthreatening guise of "first visitors."[7] Just like tourists today, it seems these ancient nature lovers did not really use or occupy future park areas. Apparently, they possessed an innate appreciation for wilderness as a place where, to paraphrase the 1964 Wilderness Act, humans are visitors who do not remain.[8] Amazingly, if we follow this reasoning to its logical extreme, the park service has managed to protect the only areas on the North American continent that Indians did not use on a regular basis.

Of course, this all sounds absurd, but scholars and park officials alike have long asserted that native peoples avoided national park areas because these places were not conducive to use or occupation.[9] Yet nothing could be further from the truth. The foothills, mountains, and canyons of most western parks provided shelter from winter storms and summer heat, sustained seasonal herds of important game animals, and served as the locale for tribal gatherings and important religious celebrations. In short, native peoples made extensive use of these areas—often well into the twentieth century. To the degree that such practices ceased, the lack of use was the result of policies to keep Indians away from these areas. Unfortunately, subsequent denials of native claims on parks have served only to perpetuate the legacy of native dispossession.

Besides taking issue with park histories that ignore the presence of Indians, this book also examines the changing importance of Yellowstone, Glacier, and Yosemite national parks for several different native groups. The people with the strongest connections to these parks include the Crow, Shoshone, and Bannock in Yellowstone; the Blackfeet in Glacier; and the Yosemite Indians in Yosemite. All have very distinct traditions, and the native presence in one park hardly resembled that in another. Blackfeet use of Glacier National Park, for instance, differed markedly from that of the Indians in Yosemite. Likewise, native use of both these places changed considerably from the middle of the nineteenth century to the 1930s, as had the lifeways of the people who lived in these areas. At Yellowstone, several groups could occupy the same general area at the same time but often for very different purposes. At all of these parks and within each Indian community, a great deal of task differentiation by gender and age group also determined the seasonal or historical importance of a particular area. During the early reservation era, for instance, male hunters accounted for most Blackfeet use of the Glacier area in summer and fall. In earlier and later periods, however, women used the area more frequently, particularly in spring and early summer, when they gathered important food and medicinal plants.

Despite their often pronounced differences, the Crow, Shoshone, Bannock, Blackfeet, and Yosemite all shared important similarities: each utilized or lived within a national park at the time of its establishment, all were affected by federal efforts to preserve certain western landscapes, none ever fully relinquished their claims to these areas in a treaty with the United States, and each park remained important to these different groups because it was large enough to protect and sustain numerous resources. While these native groups all present a powerful challenge to long-held ideas about pristine wilderness and its preservation, their use of national park lands also sheds new light on the continuing but changing significance of such areas for many Indian peoples. During the late nineteenth and early twentieth centuries, a series of harsh assimilationist programs required their adoption of new land use practices both on and off the reservation and threatened to destroy tribal societies.[10] In the midst of these profound changes, many of the places associated with older cultural practices took on new meanings or acquired new importance. Consequently, access to national park lands became a crucial aspect of native efforts to both ensure cultural survival and assert threatened treaty rights.

By examining the political, spiritual, and social importance of national park areas to different native groups, I explore the same issues that inform current American Indian concerns about the management of Devil's Tower National Monument, the industrial and commercial development of the Black Hills, and the sanctity of ancient religious sites on public lands throughout the West. This book is not just about the sacredness of certain places, however. It also addresses the rights and needs of native peoples to maintain their cultural distinctiveness through the exercise of treaty rights and the practice of certain skills that can take place only within a large national park. Recent concerns about hunting or gathering traditional food and medicinal plants on protected lands are frequently associated with a new round of cultural revivalism among various Indian groups, but these activities are rooted in a century of "illegal" and extralegal use of such areas. While these actions have presented a constant challenge to the idealization of pristine, uninhabited landscapes, they also contributed another "cultural construction" of wilderness—in this case, one in which concerns about subsistence gave way to concerns about cultural persistence and political sovereignty.

To show the ways that native peoples and wilderness enthusiasts have valued and shaped three of the nation's oldest and most revered parks, I have chosen to present this study in four parts. The first two chapters examine the ideas and historical processes that eventually led to the almost simultaneous development of national parks and Indian reservations in the years following the Civil War. The subsequent discussions of Yellowstone, Glacier, and Yosemite focus on the native histories of each park and the ways that preservationist ideals shaped policies of Indian removal or exclusion. Although the early history of Yellowstone demonstrates a close connection between the evolution of national parks and that of Indian reservations, Glacier presents a maturation of these two related but conflicting institutions. Both Yellowstone and Glacier served as important models for later preservationist efforts, and each one indirectly inspired the policies of Indian removal developed at Yosemite in the 1930s. Native residence in Yosemite Valley developed from a number of unique conditions, but park officials sought to emulate conditions in other national parks once the presence of Indians proved too exceptional. Although Indian removal has largely made these parks into American symbols of wilderness, continued restrictions on native use of park lands remain an important point of contention between many Indian tribes and the Department of the Interior. For that reason, I end this study with a chapter that connects the histories of these three parks with current concerns about nature preserves and indigenous rights throughout the United States.

As America's holiest shrines, national parks reflect a whole spectrum of ideas about nation, culture, and even natural origins. The examples of Yellowstone, Glacier, and Yosemite national parks clearly illustrate these tendencies. The early history of these parks also demonstrates how different groups, with opposing ideas about the importance of a particular place, often expressed their concerns in remarkably similar terms—and were often motivated by similar needs and historical processes. While culturally distinct and with radically different ideas about wilderness and place, Indians and non-Indians have both looked on national parks as crucial to their political, cultural, and even spiritual identity. So far, this similarity has provided only the common ground on which to base a series of profound disagreements. If anything, national parks serve as a microcosm for the history of conflict and misunderstanding that has long characterized the unequal relations between the United States and native peoples. As common ground, however, national parks might also provide an important arena of understanding and resolution—and it is toward that goal that I devote this book.

I

LOOKING BACKWARD AND WESTWARD

The "Indian Wilderness" in the Antebellum Era

> The [Indian] nations will continue to wander over those plains, and the wild animals, the elk, the buffaloe, will long be found there; for until our country becomes supercharged with population, there is scarcely any probability of settlers venturing far into these regions. A different mode of life, habits altogether new, would have to be developed.
>
> Henry M. Brackenridge, 1817[1]

TRAVELING SLOWLY UP THE Missouri River in the summer of 1832, George Catlin constantly marveled at the grand vistas stretching off to the horizon in every direction. Choked with snags and thick with mud from the spring floods, the brown waters of the Missouri cut a broad ribbon through an endless expanse of green plains, white clouds, and blue sky. For Catlin, this was "fairy land" and he never tired of "indulging [his] eyes in the boundless pleasure of roaming over the thousand hills, and bluffs, and dales, and ravines."[2] Having come west to "immerse [himself] in the Indian Country [and produce] a literal and graphic delineation of the . . . manners, customs, and character of an interesting race of people," the beauty of the landscape unfolding before him only strengthened his resolve to visit every tribe on the continent. As much as he gloried in the scenery of the upper Missouri, he could also foresee the future demise of the vast herds of buffalo, elk, and antelope that scattered in all directions whenever the steamboat on which he traveled pushed close to shore. Consequently, his enthusiasm about the landscape and the people who called it home was tempered by a sense of desperation to describe and paint all that he saw before it fell to the "desolating hands of cultivating man."[3]

Catlin had a keen sense that his success as a painter would derive from the ephemeral nature of his subject, but he did not relish the underpinnings of his future fame. Instead, he hoped that some portion of the region over which he traveled might be set off from development to inspire future generations of painters and travelers as they became "further . . . isolated from . . . pristine wildness and beauty." In what many scholars have identified as the first expression of the national park idea, Catlin proposed that "some great protecting policy of government" preserve a large expanse of land in all "its pristine beauty and wildness . . . where the world could see for ages to come, the native Indian in his classic attire, galloping his horse . . . amid the fleeting herds of elks and buffaloes." Such a "magnificent" area, he exclaimed, would be a "*nation's Park* containing man and beast, in all the wild and freshness of their nature's beauty!"[4]

The great stands of cottonwood that once crowded the Missouri's banks have long since been thinned by the very steamboats that carried travelers like Catlin. Likewise, the rolling plains have given way to farms, ranches, and small cities like Bismarck and Pierre, while long stretches of the river itself have become artificial lakes behind Gavins Point, Fort Randall, Oahe, and Garrison dams. Nevertheless, some of what impressed Catlin in the 1830s seems almost unchanged, and preserved areas like Theodore Roosevelt National Park in North Dakota serve as small replicas of the places Catlin wrote about and painted. For Catlin and his contemporaries, however, the protected scenic areas that might inspire a traveler today would seem horribly empty. Indeed, Catlin had traveled to the plains to experience what was then called an "Indian wilderness," and he would no doubt describe these areas today as "vast and idle waste[s], unstocked and unpeopled for ages."[5]

Environmentalists, park officials, and historians have long regarded Catlin as the patriarch of an intellectual genealogy that includes Henry David Thoreau, John Muir, Aldo Leopold, and the environmental movement of the past three decades. In doing so, they have largely ignored the fact that Catlin's conception of a wilderness preserve included the presence of Indians; they found, instead, only those elements that reflected on later preservation efforts. Scholars who acknowledge Catlin's desire to incorporate native peoples within a national park have generally dismissed it as something unique to his own particular interests. For them, Catlin is a man out of time: His ideas about national parks somehow foreshadowed twentieth-century concerns and policies regarding wilderness preservation; his concern for Indians, however, was either anachronistic or simply aberrant.[6]

While the devotion of his entire adult life to preserving and recording an "Indian wilderness" may have marked him off from his fellows, Catlin differed from his contemporaries only in the strength of his convictions, not in the substance of his ideas. Antebellum Americans did not conceive of wilderness and Indians as separate; indeed, the felicity with which we can speak of one *and* the other, wilderness *and* Indians, would not have been so readily conceivable in Catlin's age. Since the colonial era, Anglo-American conceptions of native peoples and wilderness had operated within the framework of a self-reciprocating maxim: forests were wild because Indians and beasts lived there, and Indians were wild because they lived in the forests. The majority of antebellum Americans viewed this "Indian wilderness" as an obstacle to progress, but those who expressed concern about the destruction of certain landscapes invariably took an interest in the welfare of the people who lived there.[7] Far from being an anomalous advocate for the protection of wild lands and native peoples, Catlin reflected contemporary ideas about both. Furthermore, his proposal for a "nation's Park" fit within a more widespread lament about the destruction of indigenous homelands that western expansion entailed.[8]

In some respects, Catlin should not be associated with national park history because his proposal hardly resembles any of the parks established later in the century. This is not to say that his ideas were somehow better. Catlin's vision of "classic" Indians grossly ignored the cultural dynamism of native societies, and his park would have created a monstrous combination of outdoor museum, human zoo, and wild animal park. Nevertheless, his ideas should not be dismissed as mere historical curiosities. To understand why his proposal for a national park was superseded by the idealization of uninhabited landscapes in the late nineteenth century, we must first situate it within the artistic, social, and political trends that shaped antebellum America. Doing so will not only provide a clearer understanding of early preservationist thinking but also allow for better recognition of the changing conditions that reshaped American ideas about wilderness and Indians at midcentury. As Henry Brackenridge predicted some fifteen years before Catlin's journey up the Missouri, "different mode[s] of life [and] habits altogether new" would transform American perceptions of the landscapes and peoples of the West.[9] It was these new modes and habits and the policies they engendered that ultimately led to the creation of the first reservations and national parks later in the century.

American Romanticism and the "Indian Wilderness"

Catlin's view of wilderness reflected the romantic ideals that had defined Western intellectual thought since the eighteenth century. In large part a reaction to both Continental rationalism and British empiricism, romanticism exalted intuition and personal experience over formalism and scientific precision. Celebrating the individual's soul—the "egotistical sublime," as the poet John Keats put it—romantics often denigrated urban life and turned to wild nature for inspiration. Consequently, wilderness not only offered an escape from society but also provided the ideal setting for romantic individuals "to exercise the cult" they made of their own souls.[10] Ralph Waldo Emerson expressed all of these sentiments in 1836 when he implored his countrymen to find "in the wilderness . . . something more dear and connate than in streets or villages." There they would experience "an apparition of God" and find "the organ through which the universal spirit speaks to the individual, and strives to lead back the individual to it."[11]

Because wild landscapes provided the most direct means for experiencing the Divine, romantics also found in the idea of the "natural man" a perfect expression of humanity. As the "children of Nature," the Indians of North America seemed to live free of the oppressive conditions that interminably plagued civilized societies.[12] Such ideas had flourished in Europe since the middle of the eighteenth century, but they did not have a strong impact on intellectual thought in the United States until the 1810s. Even then, American romantics generally regarded the few Indians still remaining in the East as remnants of a race long degraded and debauched by its contact with "civilization." Truly "noble" Indians either lived in the distant past, when America was yet "unspoiled," or roamed the distant lands beyond the Mississippi River.

With its emphasis on intuition and personal experience, romanticism had a profound impact on late-eighteenth- and early-nineteenth-century arts and letters on both sides of the Atlantic. The importance of natural beauty and the primitive—and the close association of both with the American landscape—caused the movement to take a decidedly different turn in the United States. More than a product of abundant natural scenery, however, a distinctly American romanticism grew out of the nationalistic fervor that followed the War of 1812. The idea of wilderness functioned as an important tool for patriotic apologists who felt compelled to refute European claims that the North American landscape was fundamentally flawed because it lacked ancient historical associations and refined pastoral landscapes. What American scenery lacked in European qualities, they argued, it more than compensated with an abundance of wilderness. As the painter Thomas Cole noted in 1833, "The most distinctive, and perhaps the most impressive, characteristic of American scenery [was] its wildness."[13] Such a strong identification with wilderness was hardly unique to Cole. He was, however, a founding member of the Hudson River School, the most influential group of American landscape painters in the first half of the nineteenth century, and his views had a powerful effect on American arts at this time.

Thomas Cole's own paintings demonstrated that one of the most distinctly American aspects of this "wildness" was the presence of native peoples within a "natural" landscape. No vision of the historical eastern wilderness was complete without reference to Indians, and Cole often inserted them into landscapes that had long since become "sterile and civilized." He also used images of Indians to arouse a sense of nostalgia and pity in order to give romantic poignancy to a scene, an effect he achieved in nearly all of his most important American landscape paintings, including *The Falls of Kaaterskill* (1826), *View on Lake Winniseeogie* (1828), *Distant View of Niagara Falls* (1830), *View of Shroon Mountain* (1838), and *Indian Pass-Tahawus* (1847). In *American Lake Scene* (1844), which depicts a series of small islands on a calm lake, Cole placed in the foreground a reclining Indian contemplating the tranquil scene. While Cole lavished most of his attention and skill on the landscape and not the small figure, the idea that the Indian appreciated the scene more deeply and more completely than the painter or the viewer defined the mood of the painting. Assuming the pose of a romantic poet or a tragic and pensive figure from classical antiquity, Cole's Indian hardly represented historical reality. Nevertheless, his presence in the scene was wholly consistent with romantic notions of the once noble but ultimately doomed savages of the seventeenth and eighteenth centuries.

The idea of wilderness also figured prominently in American letters during the antebellum era, and many writers conflated the nation's political and cultural identity with the aboriginal landscape. Like their counterparts in the visual arts, poets, essayists, and novelists self-consciously based their work on American subjects in an effort to create a national aesthetic. The first American authors to gain both national and international fame came to prominence in the 1820s and 1830s, and all focused on some aspect of Cole's "wildness." Indeed, almost the entire canon of early-nineteenth-century American literature consists of authors who, along with Ralph Waldo Emerson, insisted, "we have listened too long to the courtly muse of Europe" and must turn instead to the American landscape for inspiration.[14] Perhaps as a result of Emerson's exhortation, the works of Washington Irving, Nathaniel Hawthorne, James Fenimore Cooper, Henry Wadsworth Longfellow, and Herman Melville all focused on American subjects, and each author ruminated at great length on some aspect of the historical Indian wilderness in his most famous works.

Though outside the canon of American arts and letters, upper-middle-class women in the Northeast not only shared the aesthetic and nationalistic concerns of their male counterparts but also were largely responsible for the dissemination of these ideas through essays and poems in nationally distributed journals. Poets like Lydia Sigourney and Lucretia Davidson were widely read, and their poems about "the beautiful homes of the western men" or "the realm of Nature . . . [and] Nature's lawless child" were collected and reprinted in numerous editions.[15] Such ideas were repeated in the novels of Catharine Sedgewick, whose enormously popular *Hope Leslie* (1827) told of the romantic adventures that befell colonial settlers and their encounters with Indians. As the primary readers of early-nineteenth-century novels, women also determined many of the popular trends in American literature, and their literary tastes inspired the long slew of stories and novels about life among wild Indians that flooded the American market in the 1830s and 1840s.[16]

The fascination with peculiarly American themes and subjects was not limited to an elite circle of men and women in and around Boston and New York City, and the preoccupation with wildness reached far beyond their narrow social enclaves. As the literary historian Cecilia Tichi has noted, ideas about the Indian wilderness bordered on a "cultural obsessiveness" that reached across regional lines and "broke boundaries of genre, caste, and philosophical persuasion."[17] Though he was a defender of old republican virtues and a child of wealth and privilege, perhaps no author better understood the popular fascination with Indians and the frontier than James Fenimore Cooper.[18] In the Leather Stocking tales, a series of five novels published between 1823 and 1841, Cooper invented his most popular hero, Natty Bumpo. Embodying the tension between civilization and wild nature, Natty preferred the company of Indians in the wilderness over the restraints and moral debauchery of frontier settlements. Set during the Revolutionary War and the first decades of the Republic, the novels celebrated a wilderness past and lamented its recent destruction. To some degree, Cooper's novels assented to the methods by which "civilization" would eventually eclipse all of "primitive America," but he always tinged his narrative with a sense of guilt about "the forward march of progress."[19]

The fascination with wild America in the antebellum era and the profound ambivalence that many felt about the destruction of native societies did not stem entirely from romantic sensibilities. In many respects, ideas about the Indian wilderness reflected a growing sense of dissatisfaction with American politics and society. As nascent industrial and urban growth, increased immigration, and bitter political campaigns altered established patterns of work and community, public opinion often reflected a pervasive sense of national uncertainty and self-criticism. Furthermore, the growing rift between North and South, the persistence of slavery, and increasingly pronounced divisions between ethnic and religious groups undermined any sense of national unity and deflated the egalitarian rhetoric of political leaders. Together, these profound changes inspired a number of religious and secular reform movements to purify American society, and public debate often degenerated into a cacophony of local and national criticism. Not surprisingly, the Indian wilderness proved an ideal foil for social critics who used it as a corrective symbol of all that was wrong with America.[20]

Despite widespread criticism, a basic optimism characterized the antebellum era and actually provided the main impetus for most reform groups. As Alexis de Tocqueville observed in 1831, Americans regarded their society as being in a constant "state of improvement in which nothing is, or ought to be, permanent." [21] In other words, Americans remained ever critical of the present and always hopeful of the future. Such attitudes allowed many to bemoan what Thomas Cole called the routine desecration of "Nature's beauty . . . by what is called improvement" and yet accept it as a necessary part of "the road society has to travel." However lamentable the side effects of national growth and expansion, Americans would have to trust they would eventually "find refinement in the end."[22]

Indian Territory

Such ambivalence about the past and optimism for the future had a profound effect on government policy toward native peoples in the first half of the nineteenth century. Almost since the beginning of the Republic, government officials had struggled to develop an acceptable method for achieving what they referred to as "expansion with honor"—that is, how to incorporate tribal territories into the United States without belligerently undermining native societies. In theory, there were only two solutions to this moral quandary: assimilation or removal. While both required force or the threat of force, each had the peculiar virtue of transforming Indian lands into American farms and towns. On the face of things, assimilation was more "honorable" than outright dispossession, but few Americans accepted the possibility that a "Red Man" could become a fellow citizen of the United States. By contrast, distant relocation beyond the frontier seemed to hold the promise of a happy convergence of interests: settlers and speculators could buy land; missionaries could set up permanent missions among Indians without fearing the corrupting influence of nearby white communities; native groups would have an opportunity to incorporate the virtues of civilization at their own pace and, as they did so, have a positive influence on the more "savage" tribes of the eastern plains. Ultimately, removal would seem an ideal panacea for America's chronic "Indian problems," and its visionary appeal would supersede all arguments to the contrary.[23]

Few, if any, native people harbored sanguine views about their removal to the West, and none ever took much comfort in the ambivalent sympathies of artists and writers. For the tribes that attempted to remain in the eastern United States, the pressure of removal policies brought great divisions within each community. Some factions resorted to armed conflict with the United States, as in the Black Hawk War of 1832 that pitted Sauk and Mesquakie warriors against the U.S. Army and the Illinois Militia, or in the Seminole War that lasted from 1834 to 1842. The more famous Cherokee did not take up arms against the United States but instead brought their case against the government to the Supreme Court. They ultimately failed in their efforts to stave off removal, but a small number of Cherokee managed to remain in their Appalachian homeland. Far more perished between 1838 and 1839, however, when at least four thousand individuals died of starvation and exposure on the infamous Trail of Tears.[24]

The relocation of several native groups from the former Northwest Territory and the Southeast sharpened American perceptions of Indians and wilderness in a number of important ways. First, the conflicts generated by removal strengthened ideas about Indians as incapable of living in close proximity to white settlements. Perhaps just as significant, the process of removal also involved the creation of an official Indian Territory. Although the administrative boundaries of this area were eventually limited to present-day Oklahoma, the term Indian Territory broadly applied to all lands north of the Missouri state boundary and west of the Mississippi River, and occasionally referred to parts of northern Michigan. Marked off from the rest of the nation by a so-called Permanent Indian Frontier of strategically located forts, Indian Territory became a place of both the future and the past: here would be the place where Indians could develop the habits of civilized people and eventually become incorporated into the United States; here too was the place where, as James Fenimore Cooper phrased it, those interested in seeing "real" Indians would have to travel if they wished to see them "in any of [their original] savage grandeur."[25]

Like Cooper's pronouncement, George Catlin's decision to travel in Indian Territory reflected the romantic hyperbole that characterized American arts and letters at the time. Nevertheless, his proposal for a "nation's Park" also fit within the larger context of antebellum Indian policies. Although Catlin certainly would have opposed the forced removal of Indians to the West, the creation of a semiformal Indian Territory allowed him to consider a policy that might prevent the further dissolution of some native societies. In this last respect, his views partially coincided with the architects of federal Indian policy, who argued that a clearly defined Indian Territory would allow the government to better protect native communities from white encroachment. The purpose of such protection was to ensure a more orderly process of assimilation, but Catlin hoped that some effort might be expended to protect the cultural autonomy of more distant tribes. Ultimately, Catlin's proposal represented a significant departure from the ambivalent hope and resignation that characterized antebellum society, and his concern for the lands and peoples he encountered in the West would soon find echoes in the experiences of others who followed in his footsteps.[26]

Of course, the "pristine wildness" that so exhilarated Catlin in the 1830s was the product of recent developments, and several of the tribes he encountered on his journeys had just arrived from their homes east of the Mississippi. Their arrival affected indigenous horticultural groups like the Pawnee, who were already locked in a struggle with Lakota and Dakota nomads that had migrated down from the western Great Lakes region over the previous three generations. By the 1830s, some of the more sedentary peoples had already abandoned their villages and adopted a form of equestrian nomadism that allowed them to compete with the powerful Sioux. Within a few years of their arrival, a number of the emigrant tribes from the East also embraced the life of equestrian nomads in order to hold their own against the mounted hunters and warriors of the plains.[27] In short, the "native Indian . . . galloping his horse" was in the midst of revolutionary social change, and the "nation's Park" that Catlin proposed for the benefit of future generations of Americans was a cluttered arena of cultural contest and transformation.

Whether ancient residents like the Pawnee, more recently established groups like the Lakota, or brand-new immigrants from east of the Mississippi River, none of the native peoples whom Catlin met would have considered their homeland as wild. For the Sauk leader Keokuk, the land beyond the "great river" was a country that his people scarcely knew. But it would be a "new home," where "we will build our wigwams . . . [and] hope the Great Spirit will smile upon us."[28] For Lakota hunters and traders, the upper Missouri country provided a number of important resources for trade with white society and other native communities. Those groups who had resided in the region since time out of memory had a different sense about belonging to the lands they occupied; for the Ponca, the very soil on which they walked was the stuff from which their creator had made them. In every case, as the Omaha anthropologist Francis La Flesche recalled when describing his childhood on the eastern plains, the area was not a "wilderness." Indeed, to all the people of the region it was "clearly defined," and all "knew the boundaries of tribal lands; . . . every stream, the contour of every hill, and each peculiar feature of the landscape had its tradition. It was our home, the scene of our history, and we loved it as our country."[29]

Looking Toward the Western Wilds

Though out of step with native views of their homelands and certainly no more plausible than government promises to "forever secure and guarantee" these western lands to the Indians who lived there, Catlin's vision did reflect some of the reality of federal Indian policy in the 1830s and 1840s. However temporary, the "permanent" Indian frontier lasted long enough to allow a generation of artists, writers, and travelers to experience an Indian wilderness that confirmed all their romantic expectations.[30] Furthermore, western travel was made all the easier by the establishment of military outposts along the semiofficial frontier. Ostensibly designed to protect emigrant tribes from attack by indigenous groups and maintain order along the frontier, the forts also served as places of trade with western tribes and as staging grounds for upriver trappers.[31] In a very real sense, then, the maintenance of a distinct Indian Territory made an "authentic" wilderness experience possible. Ultimately, such experiences would inspire a number of prominent Americans to share Catlin's desire that some part of this region might escape the earlier fate of the eastern wilds.

In the same year that Catlin made his voyage up the Missouri River, Washington Irving returned to the United States after living abroad for seventeen years. Hoping to begin his career anew and charged with a desire to write on distinctly American subjects, he quickly made plans to visit the Indian Territory. As he explained in a letter to his brother, the prospects of such a journey were "too tempting to be resisted: I should have an opportunity of seeing the remnants of those great Indian tribes . . . I should see those fine countries of the 'far west,' while still in a state of pristine wildness, and behold herds of buffaloes scouring their native prairies." In this "tour of the prairies," as the book he later published about his travels would be titled, Irving recognized the opportunity to write on a subject that would celebrate a uniquely American condition. More important, he also saw an opportunity to record a way of life and scenery that seemed fated to "vanish."[32]

While Irving's introduction to the prairies did not lead him immediately to call for the establishment of a wilderness preserve, over the next few years he would come closer to this view in several of his most popular works. In *The Adventures of Captain Bonneville* (1837), Irving drew on his own experiences as well as Bonneville's journal to produce an adventurous story about the captain's military expeditions in the West. Irving intended *Bonneville* to preserve on the page what he termed "the romance of savage life"—the life of trappers, Indians, and wildlife. He did not simply wish to see the western wilderness preserved in print, however, and he expressed a hope that parts of the Rocky Mountains might be preserved in fact as well. Within "an immense belt of rocky mountains and volcanic plains, several hundred miles in width," he wrote in the last pages of *Bonneville*, certain places "must ever remain an irreclaimable wilderness, intervening between the abodes of civilization, and affording a last refuge to the Indian." Although the existence of such a place had more to do with the inaccessibility of the area than any governmental action, Irving hoped it would forever remain inviolate. Located near the headwaters of the Yellowstone and Snake Rivers, this "last refuge" corresponded with the area that later became Yellowstone National Park.[33]

Even more significant than Irving's "sketches of western life," the work of John James Audubon inspired a growing appreciation for the western wilderness.

Much has been written about Audubon's efforts to preserve wildlife, but scholars have paid scant attention to his concern about the demise of Native American societies. Like Catlin and Irving, Audubon's conception of wilderness and the landscapes he hoped to see preserved included native peoples. While on a trip to Labrador in the summer of 1833 to record specimens for his masterwork, *The Birds of America* (1827–1838), he repeatedly lamented the rapid destruction of the region and hoped that some "kind government" would intervene to stop its "shameful destruction." As things then stood, the destruction of deer, caribou, birdlife, and "aboriginal man" led Audubon to observe that "Nature herself seems perishing" and that there seemed to be no place left where one could go and "visit nature undisturbed."[34]

Audubon repeated these sentiments ten years later on his travels through the Ohio River valley. He noted with great sadness the changes that had been wrought on the area where, twenty years before, he had first begun his quest to paint the avian wildlife of North America. Preferring the region's previous condition to that created by its new inhabitants, Audubon recalled the "grandeur and beauty" that once characterized the river and "the dense and lofty summits of the forest . . . that everywhere spread along the hills, and overhung the margins of the streams." But this recollection lacked any of the sweetness of nostalgia. All had been destroyed by "the axe of the settler" in the intervening years; as he noted later, even the remnants of the forest would soon be lost to the "greedy mills" forever. Just as poignantly, he lamented that there were "no longer any Aborigines . . . to be found there, [nor] the vast herds of elks, deer and buffaloes which once pastured on these hills and in these valleys." In short, it was a horrible tragedy that "this grand portion of [the] Union" had not been left "in a state of nature"—with Indians, forests, and wildlife.[35]

Audubon made his trip down the Ohio en route to joining an expedition up the Missouri River. Though fifty-eight years old, he could not resist the opportunity to continue his work in the West. Along the Missouri he found scenery that reminded him of the Ohio River country some twenty years before, and he delighted in the abundance of wildlife and the grand expanse of the prairies and plains. Just twelve years after Catlin's trip up the Missouri, he already saw the effects of white settlements and commercial hunters and predicted the region would soon suffer the fate of the Ohio Valley. Though he marveled at the immense herds of buffalo, Audubon clearly recognized that their numbers were diminishing. As he noted in his journal, "before many years the Buffalo, like the Great Auk, will have disappeared"; he added that "surely this should not be permitted." Furthermore, many of the populous villages that Catlin had visited just a few years before had been decimated by disease, and those tribes that still lived along the Missouri frequently elicited pity from Audubon, their reduced condition a reflection of the impending "doom" that would soon descend upon the whole region.[36]

Educated gentlemen adventurers were not the only ones who traveled to the West, nor were they the only ones to infuse it with romantic qualities. While trapping on the upper Yellowstone River in the fall of 1834, Osborne Russell came

The Indian wilderness. George Catlin, *Mouth of the Platte River, 900 Miles above St. Louis*, 1832. Catlin wrote of the area that is now Omaha, Nebraska: "The mouth of the Platte, is a beautiful scene, and no doubt will be the site of a large and flourishing town, soon after Indian titles shall have been extinguished to the lands in these regions. . . ." Though Catlin sought out the "wilder" tribes who lived farther up the Missouri River, the lower stretches of the Platte served as the aesthetic and political model of Indian Territory for most western travelers. Home to indigenous, nomadic, and immigrant groups, the area would eventually become the gateway for overland migrants to Colorado, the Oregon Territory, and California. (Courtesy of the National Museum of Art, Smithsonian Institution, gift of Mrs. Joseph Harrison Jr.)

across some twenty or thirty "perfectly contented and happy" Shoshone encamped in an especially beautiful alpine valley. As Russell noted in his rambling prose, "I almost wished I could spend the rest of my days in a place like this where happiness and contentment seemed to reign in wild romantic splendor surrounded by majestic battlements which seemed to support the heavens and shut out all hostile intruders." A year later, he returned to the same valley and again could not refrain from commenting on the special qualities that seemed to infuse the idyllic lives of the Shoshone who lived there. Of all the places that Russell explored and trapped, none moved him as deeply as this "Secluded Valley," and the presence of the Shoshone as much as anything else made it a place time could "never efface from memory." If Russell could have visited this same valley later in the century, when it became part of Yellowstone National Park, he certainly would have recognized its scenery. The absence of the Shoshone would have marred its "wild romantic splendor," however, and he probably would not have considered the area a wilderness at all.[37]

Few Americans had an opportunity to travel west in the first half of the nineteenth century, and they could experience the western wilds only vicariously through the writings of more fortunate travelers. Still more flocked to Catlin's exhibitions as they toured the East, admired popular lithographs based on his paintings and others' works, or read the novels of Cooper and the poetry of countless romantic poets. Nevertheless, an appreciation for the Indian wilderness was manifest in the local concerns of easterners of all social classes. In New Hampshire in 1853, for instance, five hundred working men and women petitioned the Amoskeag Manufacturing Company not to cut down a stately elm tree during the construction of an additional mill. It was "a beautiful and goodly tree," they proclaimed, belonging to the time "when the yell of the red man and the scream of the eagle were alone heard on the banks of the Merrimack." The tree "belonged" in Amoskeag, which could not be said of more "giant edifices filled with the buzz of busy and well remunerated machinery," and every day the workers looked on the giant elm they felt "a connecting link between the past and the present." The mill workers could not travel to the West, but they shared the romantic concern about its destruction and could not bear to have what little of the Indian wilderness that remained in their lives cut out from under them.[38]

The Idea of Wilderness at Midcentury

As Americans of various backgrounds expressed a growing concern about the price of industrial progress, many took comfort in the fact that some portions of the precolonial landscape remained undeveloped in the East. In particular, stretches of uncut forest in the Adirondack and Allegheny Mountains attracted a growing number of outdoor enthusiasts from the cities of the eastern seaboard. Nevertheless, a clear distinction was made between the western wilds and the "pristine" mountain districts of Pennsylvania, New York, New England, and North Carolina. As someone who knew all of these areas, Charles Lanman was able to make fine distinctions between "actual" wilderness and less "pure" forms of nature. An editor, librarian, essayist, and landscape painter, Lanman started his travels in the 1830s with a trip to Maine, and over the course of the next two decades he journeyed throughout northern Michigan, the Alleghenies, and the Adirondacks. A self-described "lover and defender of the Indian race," Lanman shared the sentiments of many other young adventurers and considered native peoples to be an integral part of the wilderness. In describing Sault Sainte Marie during a trip to the northern Great Lakes in 1846, for instance, he wrote that it lay "in the bosom of a mountainous land, where the red man yet reigns in his native freedom. Excepting an occasional picketed fort or trading house, it is yet a *perfect wilderness*."[39]

On a trip to the Adirondacks in 1853, however, Lanman provided a contrast to the "perfect wilderness" around Lake Superior. While touring Lake George in upstate New York, he noted how the western shore had long been converted into farmland while the eastern shore of the lake was "yet a comparative wilderness." It was the absence of Indians to the east of the lake, coupled with sparse settlements, that defined the area as a "comparative wilderness." While beautiful in its own right, such an area by no means left as profound an impact on Lanman as did Sault Sainte Marie. Farther north of Lake George, however, Lanman was deeply impressed with the "alpine wilderness" around Mount Marcy, though in a profoundly different way. Because the area had "long since been abandoned by the red man, the solitude of its deep valleys and lonely lakes for the most part [was] more impressive than that of the far-off Rocky Mountains." Though contrary to both Washington Irving and Osborne Russell's ideas about the Rockies, the almost unnatural solitude of the Adirondacks would ultimately become enshrined in the first national parks.[40]

Any discussion of antebellum ideas about wilderness must close with an examination of Henry David Thoreau's philosophy.[41] Scholars generally agree that Thoreau's ideas about wilderness crystallized during his two-year stay at Walden Pond, when he broke his sojourn there to visit the Maine woods in the spring of 1846. While in Maine, he attempted to climb Mount Katahdin, but dangerous weather conditions and lack of adequate provisions sent him scrambling down for safety. After failing twice to ascend the mountain, Thoreau was shocked by the awful indifference that wild nature apparently exhibited toward humans; far from a transcendental encounter, the raw Maine wilderness provided a nightmare in which "Titanic, inhuman Nature has got [man] at disadvantage, caught . . . alone, and pilfers him of some of his divine faculty. She does not smile on him as in the plains."[42]

Thoreau's experience on the broken granite face of Katahdin shook the foundations of his understanding of the natural world, but this traumatic episode also brought forth the basic elements of his own philosophy. Forced to question the meaning of existence at the most fundamental level, in the most fundamental language, Thoreau wrote: "What is this Titan that has possession of me? Talk of mysteries!—Think of our life in Nature,—dayly to be shown matter, to come in contact with it,—rocks, trees, wind on our cheeks! the *solid* earth! the *actual* world! the *common sense! Contact! Contact! Who are we? where* are we?" Clearly, for Thoreau, the point of climbing Katahdin was not to find Emerson's "apparition of God" but to touch, taste, smell, and breathe nature itself. It was *"Contact!"* with primordial earth that allowed people to fully experience their humanity and not, as Emerson suggested, the relaxed contemplation of nature as if it were "a metaphor of the human mind."[43] Nevertheless, Thoreau did not leave Maine with a clear set of ideas, and it would take him several years to reconcile his dramatic experiences on Katahdin with his earlier wanderings in the fields and woods of eastern Massachusetts.

Some have argued that Thoreau's high estimation of Native Americans was considerably lessened by his trip to Maine. For Thoreau, the mountains in northern Maine seemed to be "a place for heathenism and superstitious rites—to be inhabited of men nearer of kin to the rocks and wild animals than we."[44] But it was just such people that Thoreau would strive to emulate in the following years; by developing what he called "Indian wisdom," he hoped to come nearer to rocks and wild animals. In *Walden*, Thoreau explained that he went to the woods because he "wished to live deliberately, to front only the essential facts of life." Such deliberate living was perhaps best expressed through simple language, and Thoreau feared that "we are in danger of forgetting the language that all things speak without metaphor."[45] As he noted elsewhere in his voluminous *Indian Notebooks* and *Journals*, Thoreau believed that Native Americans spoke this language and that their ways of life could teach him much about living deliberately. "The eloquent savage," he wrote, "uses nature as a symbol. . . . He looks around him in the woods . . . to aid his expression. His language, though more flowery[,] is less artificial." In short, "what [Indians] have a word for they have a thing for."[46]

Thoreau believed that true languages concerned only the essential facts of life. By learning about and from Indians, he expected to better understand his place in the universe and reconcile himself to the awesome powers in nature. Toward these ends he twice returned to Maine, and he traveled west to the prairies of Minnesota to make direct contact with native peoples. Ultimately, Thoreau expected that an appreciation of "Indian wisdom" would answer the most fundamental questions he asked himself on the face of Mount Katahdin: "*who are we? where* are we?"

When Thoreau made his famous statement that "in Wildness is the preservation of the World," he did not equate the protection of vast landscapes with the preservation of the world.[47] Instead, Thoreau spoke of wildness as a quality that all people should possess, a quality he felt was most clearly understood and appreciated by native peoples. Though ill health prevented Thoreau from making a more extensive tour of the northern prairies and eastern plains, like Catlin, he believed that some large tract of land should be preserved for native use. His clearest statement on this matter came in 1858 when he asked: "Why should not we . . . have our national preserves . . . in which the bear and panther, and some even of the hunter race, may still exist, and not be 'civilized off the face of the earth?' . . . Or should we, like villains, grub them all for poaching on our own national domains?"[48] While Thoreau certainly hoped that Indians would be able to continue their traditional lifeways relatively unmolested, his motivations were somewhat selfish. As the keepers of true wisdom, of wildness itself, Thoreau hoped they would become a reservoir of knowledge upon which the rest of the nation could draw.[49]

Scholars have universally praised Thoreau as the nineteenth century's most influential wilderness philosopher. Largely unread in his own lifetime, Thoreau's work had a profound impact on the thinking of leading preservationists like John Muir, and his works have continued to inspire countless outdoor enthusiasts around the world. Despite this legacy, Thoreau represented a way of thinking about wilderness that ended soon after he died in 1862. In many respects, antebellum conceptions of nature culminated in Thoreau's philosophy, and his was the last plea for the preservation of some portion of an "Indian wilderness." While his cry for *"Contact!"* certainly resonated with later generations, Thoreau's concern for Native Americans and the development of an Indian wisdom made little or no impact on Muir or his contemporaries. The Sierra Miwok that Muir encountered in the Yosemite high country, for instance, seemed "dirty," "deadly," and "lazy." Though Muir felt that if he knew the Indians in California better, he would like them better, their "uncleanliness" precluded any possibility of his acquiring such intimacy.[50] Thoreau's philosophy continues to inspire readers more than a century after his death, but his conception of what constituted wilderness and the significance of its preservation simply did not translate over to the latter decades of the nineteenth century.

2

THE WILD WEST, OR TOWARD SEPARATE ISLANDS

We did not think of the great open plains, the beautiful rolling hills, and winding streams . . . as "wild." Not until the hairy man from the east came . . . was it wild for us. When the very animals of the forest began fleeing from his approach, then it was that for us the "Wild West" began.

Luther Standing Bear
(Oglala), 1933[1]

SCARCELY A MONTH AFTER THE close of the Civil War, Samuel Bowles realized a long cherished dream to visit what he called "the Great West." The opportunity came through an invitation from his good friend, Schuyler Colfax, who, as speaker of the House of Representatives and chairman of the House Committee on Post Offices and Post Roads, proposed a tour of the future route of the transcontinental railroad. The trip west greatly advanced Bowles's career as a newspaperman, and the series of letters he wrote for the *Springfield Republican* gave national prominence to his paper. Moreover, the collected letters provided the basis for a best-selling book, *Across the Continent: A Summer's Journey to the Rocky Mountains, the Mormons, and the Pacific States, with Speaker Colfax,* which quickly made him a leading expert on the West. Although Bowles protested that his book was neither "a Diary of a personal journey; nor a Guide Book," in truth, it derived much of its popular appeal from being a combination of both. Like Speaker Colfax, he was an apologist for Manifest Destiny who linked the nation's future with the success of the transcontinental railroad, but Bowles was first and foremost a tourist who readily described "interesting and picturesque" places that would attract a host of later travelers and sightseers. As historian Anne Farrar Hyde has noted, *Across the Continent* "reads like a blueprint for every guidebook, travel account, and tourist reminiscence to appear in the first decades of transcontinental travel."[2]

In the summer of 1868, Bowles once again received an invitation from Colfax to join a large group of friends and relatives on a trip to the Colorado Rockies. He quickly accepted the offer, which afforded an opportunity to revisit the part of the country that he most enjoyed on his previous journey. For Bowles, the mountains west of Denver were the "Switzerland of America," where "the great backbone of the continent rears and rests itself [and] . . . nature sets the patterns of plain and mountain, of valley and hill, for all America; . . . here, indeed, is the center of the central life of America,—fountain of its wealth and health and beauty." The center of this American Switzerland lay in the "wide elevated Parks, lying among her double and treble folds of the continental range . . . surrounded by mountains that rise from . . . plains, green with grass, dark with groves, bright with flowers." The spiritual and scenic heart of this region—and therefore of the entire nation—was the Hot Springs Valley in Middle Park, which he and his companions visited for several "exhilarating" days in mid-August.[3]

As he later described the experience in his *The Switzerland of America*, Bowles came upon his first view of Hot Springs Valley after climbing a small hill, from which he gazed down upon "a broad, fine vision. Right and left, several miles apart, ran miniature mountain ranges,—before, six miles away, rose an abrupt gray mountain wall; just beneath it, through green meadow, ran the [Colorado] River . . . [where] a hundred white tents, like dots in the distance, showed the encampment of eight hundred Ute Indians." "In the upper farther corner," he continued, "under the hill-side, a faint mist and steam in the air located the famous Hot Springs of the Middle Park,—the whole as complete a picture of broad, open plain, set in mountain frame, as one would dream of. It spurred our lagging spirits, and we galloped down the long plane."[4]

This description of Middle Park certainly resembles what Osborne Russell longingly referred to as his "Secluded Valley" and Washington Irving called the "last refuge [of] the Indian."[5] Bowles had not come west to see Indians, however, and he found nothing in their presence that added to his experience in the Rockies. Besides, the several bands of Mountain Ute had not gathered in the valley simply to engage in the sorts of "picturesque" and "romantic" activities that might otherwise attract a tourist like Bowles. They had come to meet with the governor of Colorado Territory about a pending treaty agreement to cede the whole of Middle Park and other lands to the United States. Although Bowles noted how much the Ute were "loth to yield control of [the Hot Springs and Middle Park] to the whites," he felt the "scheme [was] a good one" and had no qualms about their involuntary removal from the area. As he saw it, the benefits of the plan were twofold: the treaty was good for the Indians because it "moved them away from the mines and the whites" to a place where they could engage in "a pastoral and half agricultural life"; for the United States, the treaty served as the best prescription for opening up large tracts of land for mining, agriculture, and settlement. Perhaps most important to Bowles, the treaty also cleared American title to a "wedded circle of majestic hill and majestic plain" that he predicted would soon become "the pleasure ground and health home of the nation."[6]

In many respects, Samuel Bowles represented the beginning of a new movement that would lead to the creation of the West's largest and most celebrated national parks. On his previous trip through the West in 1865, he visited Yosemite and praised the recent congressional act that placed the valley and the Mariposa Big Trees under the protective authority of the California legislature.[7] Bowles hoped that Yosemite Park, as the small, state-administered reserve was then called, would serve as "an admirable example for" the preservation of Niagara Falls, a section of the Adirondacks, some portion of New England's lakes and forests, and "other objects of natural curiosity and popular interest all over the Union." To preserve such areas, he exclaimed on a later occasion, would be "a blessing to . . . all visitors . . . [and] an honor to the Nation!"[8] Nowhere impressed him as much as Middle Park, and though the area offered "no wonderful valley like Yo Semite; . . . no cataract like Niagara; no forest like those of the Sierra Nevada range, no, nor the equals, in diversified form and color and species, of those of New England or of Pennsylvania," none of these places could so greatly bless visitors and honor the nation as "these central ranges of continental mountains and these great companion parks."[9]

In much the same way that he anticipated the movement that would soon lead to the creation of the first national parks, Bowles also proved a strong advocate for the government's newly developing system of Indian reservations. Like most people who cared to think of such matters, Bowles proposed the cessation of all treaty councils and felt the government should unilaterally dictate terms to the western tribes. "We know they are not our equals," he argued, "[and] we know that our right to the soil, as a race capable of its superior improvement, is above theirs; [therefore,] let us act directly and openly our faith." "Let us say to [the Indian]," he continued, "you are our ward, our child, the victim of our destiny, ours to displace, ours to protect. We want your hunting grounds to dig gold from, to raise grain on, and you must 'move on.'" According to Bowles, the government was required to "give" the western tribes a number of small reservations of their own, but native leaders needed to understand that whenever "the march of . . . empire demands this reservation of yours, we will assign you another; but so long as we choose, this is your home, your prison, your playground." While Bowles recognized the inherent "dishonor" of such policies, he could rationalize them with a solid conviction that native peoples were doomed to "vanish." Consequently, the government's only responsibility was to feed and educate the Indian "to such elevation as he will be awakened to, and then let him die,—as die he is doing and die he must."[10]

Bowles might have added that Indians would also have to "move on" whenever Americans valued the scenic or healthful qualities of certain landscapes. Even when they did not wish to mine or farm "hunting grounds," Americans could not abide the continuance of native societies on some portion of the public domain. As Bowles put it, the Indian's "game flies before the white man; we cannot restore it to him if we would; we would not if we could; it is his destiny to die; we cannot continue to him his original, pure barbaric life; he cannot mount to that of civilization." There was nothing to do but "smooth and make decent the pathway to [the Indians'] grave." The important thing for Bowles was not to mourn what he viewed as inevitable, but to get on with the business of using and enjoying the recently vacated lands of America's "Great West."[11]

Nature Sets the Patterns for All America

Coming only a generation after George Catlin's journey up the Missouri River, the purpose of Samuel Bowles's "summer vacations" could not have contrasted more sharply with the former's "residence and travel" on the plains and prairies. Nothing reflects this difference better than each man's attitude toward the native groups he encountered. For Catlin, of course, meeting Indians was the object of his travels, and the upper Missouri country seemed a vast refuge for North America's original Indian wilderness. Bowles viewed the Far West as a realm of great symbolic, material, and recreational promise that could not be fully realized until native peoples had been rounded up on reservations and made to die a quiet death.

Though certainly a matter of individual differences, the marked contrast of their views also stemmed from the years of tremendous national change that separated their two journeys. In short, Catlin and Bowles operated within two different worlds and interpreted two very different "Wests." The latter's excursions to the Rocky Mountains and the Pacific coast more than reflected the political geography of a vastly enlarged United States; his ability to find national symbols in the spectacular landscapes of the Far West had also been shaped by the legacies of both the Mexican War and the Civil War. Likewise, his opinions about the native peoples he encountered derived from a series of dramatic shifts in federal Indian policy that had accompanied national expansion and the wars it engendered.[12]

Unlike the Louisiana Territory, which was peacefully acquired from a European power in 1803 and still only sparsely settled by Americans when Catlin began his journey up the Missouri River, the Far West was rapidly and violently incorporated into the United States in just a few years. The conquest of northern Mexico and the annexation of Oregon fulfilled what New York newspaperman John O'Sullivan had called America's "manifest destiny to overspread and . . . possess the whole of the continent for our yearly multiplying millions." In this context, the spectacular landscapes of the "New West" not only became trophies of war that glorified a new continental empire but also symbolized the nation's divine covenant with Providence to bring liberty and democracy to the shores of the Pacific and beyond.[13] Because expansionists like O'Sullivan believed the "True Title" to these lands rested on some timeless principle of geopolitical predestination, they argued that Americans had a moral and biological duty to extend the Anglo-Saxon "race" over the western half of the continent and either subjugate or extinguish the inferior "racial strains" that currently occupied these lands.[14]

The jingoistic nationalism that precipitated the Mexican War and convinced the British to withdraw their claim to the Oregon Territory in 1846 profoundly shaped the ways that Americans would perceive the landscapes and peoples of the West for several decades. Likewise, the conflation of racial, political, and geographic "destinies" with the cant of conquest effectively erased the human history of western North America and replaced it with an atemporal *natural* history that somehow prefigured the American conquest of these lands. The "discovery" of Yosemite Valley in 1851, for instance, revealed a perfect "natural monument" to the newly expanded United States—just three years after the area had been ceded by Mexico and even before the resident Indians who called it home had been temporarily driven away.[15] Similarly, the giant sequoias in the nearby Mariposa Grove imparted an instant antiquity to the United States that rivaled the ancient cultures of Europe and connected the American landscape—and thus American civilization—with a physical past that reached back to the time of ancient Rome.[16]

The acquisition of a vast expanse of territory also destroyed any earlier pretenses about a Permanent Indian Frontier along the 100th meridian. The notion that most of the land between the Rocky Mountains and the Missouri River would somehow remain Indian country for even a generation or two collapsed once growing numbers of Americans crossed this region in the 1840s and 1850s on their way to Oregon, California, and Colorado. For emigrant tribes living within the present-day boundaries of Iowa, Missouri, Kansas, and Nebraska, an invasion of permanent settlers undermined their recently established communities and forced them into another series of land cession agreements with the United States. The passage of the Kansas-Nebraska Act in 1854 effectively nullified earlier government promises to guarantee their lands in perpetuity, and within a decade nearly all would be relocated to the now official but greatly reduced Indian Territory in present-day Oklahoma. Farther west, the growing flood of settlers and migrants would overwhelm native societies and lead to more demands for Indian land. While the idea of reserving some place for these western tribes to learn the arts of American civilization persisted, any notion that Indian removal somehow benefited native groups as much as it did white settlers no longer served as a necessary apology for national expansion.[17]

Making the West Wild

With the exception of two major conflicts in the 1830s, the United States had been more or less at "peace" with Indians for nearly four decades. Between the mid-1850s and the late 1860s, however, vigilante groups, local militias, and U.S. Army troops fought countless battles with native peoples throughout the West. In response to expected conflict, the army built at least six dozen military forts west of the Mississippi, and almost all were used in campaigns against Indian communities. Maps of the western United States reflected this new construction, and policy makers, overland travelers, and even casual newspaper readers became familiar with places like Fort Bridger, Fort Laramie, and Fort Kearny. Indeed, the political geography of the West seemed to reflect a national single-mindedness toward migration and warfare against Indians. Aside from well-traveled rivers and mountain passes, many popular maps filled in the western landscape with the sites of forts, the names of hostile tribes, and the locations of famous conflicts.[18]

The resumption of the nation's long history of Indian wars had a profound effect on the way Americans perceived native peoples and the West. No longer picturesque and "noble" Indians who freely roamed through a distant region, the western tribes now lived on coveted lands within the national domain and regressed into "treacherous, blood thirsty savages." As Herman Melville phrased it in his novel *The Confidence Man: His Masquerade* (1857), "The metaphysics of Indian hating" seemed to define the core of the national psyche.[19] Although few Americans called for outright annihilation of Indians, Melville's assessment provides a fair representation of the ideas that shaped federal Indian policy in the decades preceding and following the Civil War. Whether Indian haters or religious reformers, almost everyone could agree that America's Manifest Destiny required the physical or cultural destruction of all native peoples. Without a farther West to push them toward, the best method for achieving this common goal was to relocate Indians onto a reserved portion of their homelands. Once there, it was hoped the army could more closely control their movements and reformers might "provide, in the most efficient manner, . . . for . . . the ultimate incorporation [of Indians] into the great body of [the] citizen population."[20]

As historian Richard White recently observed, "The reservation system grew like Frankenstein's monster, bolted together from the corpse of the older hope for a permanent Indian territory west of the Missouri."[21] Like Dr. Frankenstein, policy makers had little understanding of the people or conditions with which they worked, and their initial efforts generally proved disastrous for native peoples. In parts of the Far West where a flood of settlers and miners had already inundated Indian lands, federal officials worked through the mid 1850s to remove a number of tribes to more remote areas. Their hurried efforts only exacerbated a number of tense situations, and the new policy failed outright when local militia groups and overzealous army troops committed a series of brutal massacres in Texas, California, and western Oregon. But reservations more closely reflected military exigencies than humanitarian impulses, and these early setbacks could not be considered total failures. Belief in the efficacy of reservations persisted and, as Commissioner of Indian Affairs Charles E. Mix stated in 1858, government officials remained convinced that "concentrating the Indians on small reservations of land . . . [whenever] it may be necessary to displace" them was still the best method for "controlling the Indians" and teaching them "civilized occupations and pursuits."[22]

On the Great Plains and throughout the Rocky Mountains, different conditions led to different results. Because many of the plains and intermountain tribes presented a formidable challenge to unlimited American expansion, the government did not find it "necessary to displace" them until after the end of the Civil War. In three major treaty councils in the early 1850s, the United States recognized native rights to an area that extended from present-day Idaho, east to the Dakotas, and south to New Mexico. Instead of reducing tribal land holdings and restricting these people to remote reservations, treaty commissioners sought to establish peace with different native groups and guarantee rights of way for western travel.[23] Such promises would be difficult to keep as migrant parties and supply trains increasingly abused already overtaxed ecosystems. Once emigrants began to settle on recognized Indian lands, however, violent conflict soon followed. By the mid-1850s, just a few years after the United States had promised "an effective and lasting peace," the series of conflicts known as the Plains Wars had begun.[24]

Most, if not all, native groups preferred to avoid contact with the army, but war came despite many conciliatory efforts. Whether against the United States or another tribe, one of the overriding reasons for western native warfare was to secure or retain access to key resources. For tribes like the Pawnee and Crow, however, cooperation with the army could achieve similar goals by serving as a buffer against Sioux encroachments on their territory. Shoshone, Ute, and other intermountain groups managed to avoid conflict with both the United States and the powerful Sioux through increased reliance on the plants and animals of the Rocky Mountains. These strategies had achieved some success by the mid-1860s, but each depended on a series of precarious balances in a rapidly changing world. Within a few years, almost all plains and intermountain groups were in dire straits, and the American conquest looked more and more like a foregone conclusion.[25]

Despite their vulnerable condition, a number of "friendly" and "hostile" native groups still posed a significant threat to military installations and civilian settlements throughout the West. Consequently, as government officials learned at a series of treaty council meetings in 1867 and 1868, any efforts to open up new areas for American settlement would have to accommodate native concerns if the United States wanted to avoid the expense and danger of further warfare. When Blackfoot, a principal chief of the Crow spoke to a party of peace commissioners at Fort Laramie in November 1867, he made it abundantly clear that his people would end their alliance with the United States and go to war if they were confined to just "one corner of [their] territory." "How can we [continue to be peaceful with the United States] when you take our lands, promising in return so many things which you never give us?" "We are not slaves," he thundered, "and we are not dogs. . . . We want to live as we have been raised, hunting the animals of the prairie. Do not speak to us of shutting us up on reservations."[26]

Blackfoot's words echoed those of Kiowa, Comanche, Cheyenne, and Arapaho leaders, who, just a month before, refused to sign treaties with the same commissioners unless the tribes retained usufruct rights to off-reservation lands. Although these concerns ran counter to the legislation that authorized these treaty councils, the government was forced to change its easy assumptions about gaining "the consent of the Indians to remove to . . . reservations." Charged with making agreements that would "remove all causes of complaint on [the part of Indians], and at the same time establish security of person and property along the lines of railroad [then] being constructed to the Pacific," the treaty commissioners could hardly afford to ignore the common concerns of various tribal leaders. [27]In almost all of their meetings with native representatives, government officials heard similar demands and eventually agreed to recognize each tribe's right "to hunt on the unoccupied lands of the United States so long as game may be found thereon."[28]

These treaties may have guaranteed native use of lands within the vast tribal boundaries established in the early 1850s, but they also stipulated that Indians could make permanent settlements only on the smaller areas that constituted their reservations. As General William Tecumseh Sherman reported from the Fort Laramie negotiations in 1868, most policy makers believed that off-reservation rights were a "temporary" expedient to "gain time, and . . . withdraw from hostility a considerable part" of each tribe. Moreover, this condition helped save the United States a considerable amount of money because, in the short run, it allowed native people to depend on their own resources.[29] As Americans hunted and settled on these lands, the theory went, game animals would diminish and Indians would eventually be forced to commence farming. Once they had turned to agricultural pursuits, "excess" reservation lands could be sold to the government and, as Americans had long predicted, the future "civilization" of Indians might finally be assured. In short, government officials accepted native use of off-reservation lands because they believed that subsequent developments would make such rights obsolete.

While policy officials planned on restricting these tribes to even smaller areas, native leaders viewed their agreements with the United States in a very different light. By defining a reservation as a place "set apart for the absolute and undisturbed use and occupation" of a tribe, the treaty agreements seemed to guarantee that the United States would not allow its citizens to invade a reserved area of land.[30] This stipulation did not place any limits on the residence of tribal members. The retention of usufruct rights to areas outside the reservation boundaries meant that Indians would continue their customary movements; the only difference between reservation and off-reservation lands was that native leaders had agreed to share the latter with settlers and railroad builders. Most tribal representatives probably understood General Sherman's intentions, and some may have shared a parallel vision of their people's future, but none saw their treaty with the United States as an outright cession of land. Indeed, all of these treaty agreements hinged on what the Shoshone leader Washakie called "the privilege of going over the mountains to hunt where I please."[31]

Not surprisingly, American efforts to restrict off-reservation "privileges" met with strong resistance and soon contributed to the famous Indian Wars of the 1870s. While failure to peacefully settle Indians on fixed areas of land eventually cost all the lives and money that policy officials had tried to save in 1867 and 1868, it did not mean a repudiation of the government's reservation policy. Military conquest did not turn Indians into willing farmers, but it brought new restrictions on Indian movements and invariably led to further land reductions.[32] Throughout the central and northern Rocky Mountains, however, native peoples generally avoided conflict with the army and managed to exercise their treaty rights for several more years. Finding the sort of "refuge" that Washington Irving once described, Indians from reservations in Wyoming, Idaho, and western Montana continued to utilize certain mountain areas much as they had for generations. Their ability to maintain these practices soon encountered a powerful new challenge from a growing American concern for these same "unoccupied lands of the United States."

Toward Wilderness Preservation

In the years following the Civil War, the rapid exploitation of western lands seemed to confirm the nation's future destiny, as vast new regions came under the plow or yielded to the miner's pick and the lumberman's saw. Nevertheless, a growing number of travelers and social commentators began to question the aesthetic costs of western development. Few believed that concern for scenery necessarily outweighed financial considerations, but many argued that some better accommodation needed to be made between the two. Even as strong an advocate of western development as Horace Greeley cautioned his countrymen to "spare, preserve and cherish some portion of your primitive forests; for when these are cut away I apprehend they will not easily be replaced."[33] Likewise, the intrepid author and traveler Bayard Taylor, who otherwise gushed effusively about the rapid pace of "civilization" and development in the Far West, could not help but criticize the scenic "costs" of hydraulic mining and clear-cut timbering in California. Having visited the area in 1849, Taylor returned in the 1860s and described the dramatic changes wrought on the landscape: "Nature here reminds one of a princess fallen into the hands of robbers," he wrote, "who cut off her fingers for the sake of the jewels she wears."[34]

The appeal of so much "unspoiled nature" in the West and the fears about its imminent destruction reflected many of the same romantic sensibilities espoused by earlier artists and writers like Thomas Cole and James Fenimore Cooper. But such appreciation for the western wilderness had little to do with the presence or absence of native inhabitants. Instead, a new generation of patriotic aesthetes focused their attentions almost entirely on the physical geography of the Rocky Mountains and the Sierra Nevada. Vast primeval forests and picturesque Indians might once have distinguished American landscapes from those of Europe, but Americans could now boast of towering mountains, giant trees, and stupendous waterfalls that surpassed everything else in the known world. Moreover, the fact that such natural wonders lay two and three thousand miles from the eastern seaboard, yet within the boundaries of one nation, exemplified the continental scope and power of the United States.

Alfred Runte has rightly noted that Americans sought out spectacular locales because they possessed a certain "monumentalism," a quality that evoked a powerful sense of natural wonder and national pride. Moreover, such nationalistic identification with particular landscapes was a necessary precondition, if not a direct reason, for their protection from commercial development.[35] President Lincoln's 1864 signing of the Yosemite Park Act, which set aside fifteen square miles of the public domain and placed it under the protection and management of the

Divine confirmation of the nation's Manifest Destiny, 1874. Thomas Moran, *Mountain of the Holy Cross,* in *Picturesque America* II: 501. Wood engraving by J. Augustus Bogert. Perhaps no image better captures the sense of America's special covenant with the western landscape. Moran, who had not yet seen the famous mountain in Colorado, based this drawing on photographs taken by W. H. Jackson in 1873. Already famous for his paintings of Yellowstone and the Grand Canyon, Moran felt obliged to make his own pilgrimage to the site in 1875. The trip produced a large oil painting that was honored at the Centennial Exposition in Philadelphia. (This item reproduced by permission of the Huntington Library, San Marino, California.)

state of California, created an important precedent for the preservation of larger areas in the coming decades. Even more significant at the time, the scenic wonders of Yosemite Valley and the Mariposa Big Trees served as powerful symbols of national unity, just as the Civil War seemed to finally draw toward an inevitable conclusion.

Another important catalyst for the protection of a place like Yosemite stemmed from national embarrassment over the commercialization of Niagara Falls, which inspired a movement not only to preserve certain areas in the West but also to maintain them in as "natural" a condition as possible. Prior to the Mexican War, Niagara was the only natural feature that Americans could point to with any sense of national pride, but the proliferation of "museums, mills, staircases, tolls, and grog-shops" around the falls had all but destroyed their "sublimity."[36] Places like Yosemite Valley or Middle Park not only provided an expanded and improved set of national symbols but also offered an opportunity to redeem the mistakes that had occurred at Niagara. Tourists would come to these places—indeed, they should come—but a consensus was building that their experiences of "Nature's bounties" should remain as unencumbered as possible.[37]

A trip west or stories and paintings about western scenes did not simply appeal to a nationalistic passion for "monumental" nature. The other great attraction of these places stemmed from their recreational qualities, which would only increase with the completion of the transcontinental railroad. Samuel Bowles considered Middle Park to be America's answer to Switzerland and the rest of Europe, but he celebrated the area as much for its alpine scenery as for "the health and sentiment of the thin pure air of the Mountains and the Parks."[38] Likewise, his appeal for the state of New York to preserve some large tract of land in the Adirondacks had as much to do with scenery as with the healthful aspects of a trip to the mountains. Outdoor recreation was becoming something of a craze among well-healed easterners and, as one popular advocate of "camplife" and "wilderness adventure" wrote in the late 1860s, thousands of men and women, "weary of the city's din, long[ed] for a breath of mountain air and the free life by field and flood."[39] As Bowles and others predicted, these urban pressures would become more acute all across the United States, and then, as long-distance travel became less difficult, the Rocky Mountains would do for the entire nation what the Adirondacks were already doing for eastern urbanites.[40]

Although both recreational and scenic interests shaped this growing appreciation for certain landscapes, new concerns about environmental degradation made the preservation of large, undeveloped areas all the more important. In 1864, the same year that President Lincoln signed the Yosemite Park Act, George Perkins Marsh published an enormously influential work on land use and resource conservation, *Man and Nature*. Marsh warned that "the earth is fast becoming an unfit home for its noblest inhabitant, and another era of equal human crime and human improvidence . . . would reduce it to such a condition of human productiveness, of shattered surface, of climactic excess, as to threaten the extinction of humanity itself." Basing much of his argument on research he conducted in the Mediterranean and in his home state of Vermont, Marsh concluded that

America's rapid economic progress was predicated on wasteful processes that could ultimately draw the United States into the same tragic collapse that had befallen the great civilizations of antiquity. Ironically, the nation's exploitation of the very wilderness that distinguished it from the Old World would convert America into a sad replica of Europe. Marsh counseled that only rejection of unlimited economic expansion and careful conservation of natural resources could stave off certain environmental catastrophe and social collapse.[41]

Aggressive conservation programs might forestall the inevitable, but something more was needed to preserve American exceptionalism. "Nature, left undisturbed," Marsh wrote, "so fashions her territory as to give it almost unchanging permanence of form, outline, and proportion." Because the presence of humans invariably disturbed natural balances, certain areas still more or less "untrodden by man" must remain so to provide a model for "the restoration of disturbed harmonies and the material improvement of wasted and exhausted regions." Like a modern ecologist, then, Marsh advocated the preservation of large natural areas in order to demonstrate the workings of a healthy ecosystem. He believed it "a matter of the utmost importance" that some large portion of the public domain should "remain, as far as possible, in its primitive condition." While such a place would prove a great "garden for the recreation of the lover of nature," it would also be a much needed "asylum where indigenous tree, and humble plant . . . and fish and fowl and four footed beast, may dwell and perpetuate their kind, in the enjoyment of such imperfect protection as the laws of a people jealous of restraint can afford them."[42]

A visionary work that remained influential for more than four decades, *Man and Nature* inspired the environmental philosophies of leading conservationists well into the twentieth century. But Marsh also influenced policy makers and business leaders in his own time, and the contemporary importance of his work should not be underestimated.[43] Probably no one was more deeply affected by *Man and Nature* than Frederick Billings, and certainly no one was in a better position to implement Marsh's philosophy. Though he does not appear in any histories of conservation or preservation, Billings was involved in the creation of Yosemite Park in the early 1860s, and later, as a director and then president of the Northern Pacific Railroad, he played an important role in the creation and development of Yellowstone National Park. While his concern for Yellowstone reflected the financial interests of Northern Pacific, Billings also shared Marsh's belief in the necessity of a national "garden." By far the strongest testimonial of his admiration came with the purchase of the Marsh estate in Woodstock, Vermont, and his subsequent efforts to convert the property into a showcase of conservationist principles.[44]

Like Samuel Bowles, Frederick Billings represents the intersection of several strains of thought about wilderness and Indians in the years following the Civil War. Besides his lifelong interest in conservation, his support of the Yosemite Park Act reflected all the romantic sentiments that had long defined most Americans' fascination with the sublime and the beautiful in nature. Likewise, he saw in Yosemite and other scenic locales in the West a powerful, unifying symbol for a nation in the midst of Civil War. His interest in the creation of Yellowstone National Park in 1872 reflected a different but related set of concerns about economic development and outdoor recreation. In what his biographer has called "a shrewd judgment about commerce and tourism," Billings recognized that "commerce could serve the cause of conservation by bringing visitors to a site worthy of preservation."[45] Moreover, a place like Yellowstone could also serve as a symbol for Northern Pacific and an important destination for an expected increase in passenger traffic. Finally, as the leader of a major railroad, which sought to build new lines and open up new agricultural markets, Billings was an aggressive advocate for the further reduction of tribal landholdings. He even boasted that railroads could restrict Indians to their diminished reservations because the prairie fires caused by trains proved an effective, if somewhat accidental, measure for driving away the game that attracted native hunters to their recently ceded lands.[46]

Separate Islands of the Mind

Samuel Bowles and Frederick Billings demonstrate that attitudes toward the people and landscapes of the Far West underwent marked changes in a relatively short period of time, but their experiences do not articulate how an earlier appreciation for an Indian wilderness split into separate movements for the preservation of scenic areas and the confinement of Indians to reservations. Perhaps no person better exemplifies this transition in thinking about wilderness and Indians or better demonstrates how new ideas about both were linked to the same historical developments than George Belden. In the early 1850s, while still a teenager, Belden left his home in Ohio for Brownesville, Nebraska, a hamlet of log houses on the banks of the Missouri River. Just two years later, after convincing his family to come with the tide of emigrants descending upon Nebraska and join him in his new home, the young Belden became disenchanted with the changes in this frontier town. As he described it in his enormously popular autobiography, "brick houses [had begun] to appear; the buffalo, game, and Indians were gone, and I felt Brownesville was no longer my home. I burned for adventure, and when our little weekly paper was announced as a 'daily,' I knew it was time to go." Heading out to the Great Plains and the Black Hills of the Dakotas, Belden spent the next twelve years as a hunter and trapper while making his home with various plains Indian communities.[47]

Though not a gentleman traveler, Belden was by no means immune to the "romantic beauty" of the western landscape. While hunting with a group of Santee Sioux in the Big Horn country, he was deeply impressed with the area's scenery, proclaiming that "nothing could have been more pleasant than [camping and hunting] on the broad, wild prairies of the West." On another occasion he made special note of a particularly "beautiful prospect" that he looked down upon from a rise in the prairie: "Far away, winding like a huge silver serpent, ran the river, while nearby, in a shady grove, stood the village—the children at play on the green lawns not made by hands. The white sides of the teepees shone in the setting sunlight . . . [as] bright ribbons and red flags . . . fluttered from the lodge-poles, and gaudily dressed squaws and warriors walked about, or sat on the green sod under the trees." For Belden, as for many before him, this scene represented a perfect Indian wilderness, and as such it was all the more beautiful. Toward the end of his sojourn as the self-fashioned "white chief" of the plains, however, he would develop new views about these lands and the people who called them home.[48]

When the Civil War broke out, Belden joined the U.S. Army and spent the next few years warring against the people he had originally sought to live with. The significance of this change is thick with meaning, and Belden may have experienced some profound convergence of racial and national identity that was triggered by fratricidal war in the East. Or he may simply have become a "responsible" young man and decided to return to the land of daily newspapers. In either case, he evaluated native peoples in a new light and chastised them for not reconciling themselves to certain conquest. Belden did not feel any particular hostility toward his former companions, but he did believe that war was the best method for bringing about peace. More important, only military defeat could instill among Indians the same respect for "civilization" that he now espoused. With the zealotry of a new convert, Belden had become a champion of national progress who lost patience with his own romantic views about the "wild Indians of the plains." Brownesville was still growing, newer settlements were beginning to flourish, and there was no longer any reason for America to abandon "a rich, fertile, and beautiful country to a few thousand savages, who [could] make no use of it but to chase the lessening herds of buffalo and deer."[49]

As he came to new conclusions about his old companions, Belden also developed a new appreciation for their homelands. The Big Horn region that he so treasured would make excellent farm country, and he recommended that his native friends be moved to a distant reservation. Once there, "a remnant of the race" could be "preserved for posterity" and "turned gradually from their wild habits of roving, and living from day to day, to settle . . . and live as herders and farmers." While their former lands were brought under the plow, Belden hoped that "one of the greatest natural curiosities on the continent," the Big Horn Canyon, would be preserved for all the world to see. "Whatever there is of beauty in the wildest scene of nature," he wrote, "in the massive grandeur of rock, in the grace of vines and foliage, and the charm of running water, is furnished by this lonely cañon." More than a place for solitary contemplation of the sublime, the whole natural spectacle would also become an object of national pride and a resort for the seeker of "health and pleasure."[50]

While George Belden's twelve years on the plains reflected a profound change in mid-nineteenth-century attitudes toward western landscapes and the people who lived there, his views did not necessarily represent actual conditions. He may have separated Indians from wilderness in his mind, but several tribes continued to use Big Horn Canyon, and the area never developed into the famed resort he once envisioned. Likewise, he shared with most Americans a belief that Indians would soon "vanish," but none of the tribes he encountered gave any indication of quietly disappearing. Indeed, his desire that some part of the national domain might be protected at a time when Indians were supposed to be vanishing seems ironic when the United States disposed of its public lands far more rapidly than native populations declined.[51] On the face of things, it would seem that Belden was entirely wrong about the people and places he described. But ideas can shape reality, no matter how poorly they might reflect actual events, and such would be the case with the powerful convictions that Americans held about wilderness and Indians in the decades following the Civil War.

More than Big Horn Canyon, Middle Park, or any other part of the American West, the spectacular scenery on the upper reaches of the Yellowstone River provided "a realm of mighty marvels" that seemed the realization of America's "wildest fantasies."[52] The geyser basins, waterfalls, canyons, peaks, lakes, and forests satisfied the deepest yearnings of cultural nationalists, who found in Yellowstone an unparalleled assortment of natural monuments to the power and grandeur of the United States. In the midst of a growing desire to visit and protect such places in their "wild" condition, a movement quickly developed to convert the area into a "great public park for all time."[53] During an era of unprecedented federal activism, Congress proved very receptive to these ideas and soon drew up legislation that provided for "the preservation" of Yellowstone in its "natural condition."[54] None of the debates over the park bill even mentioned Indians, however, except to note that none lived in the designated area. Based as much on ignorance as wishful thinking, the failure to acknowledge that native peoples extensively used the Yellowstone basin would soon prove the first great challenge to the national park ideal.[55]

Passed into law on March 1, 1872, the Yellowstone Park Act removed more than two million acres of the public domain from "settlement, occupancy, or sale." In doing so, Congress inadvertently protected the "unoccupied lands" where several native groups exercised their off-reservation treaty rights. Hardly a resurrection of George Catlin's old proposal for a "nation's Park," this oversight soon proved a matter of great consternation to park officials. In an effort to correct the situation, they eventually collaborated with the Indian Service, the military, and the federal judiciary to effectively exclude Indians from Yellowstone. While the legacy of these combined efforts demonstrate that wilderness preservation is predicated on native dispossession, several indigenous groups would present a strong challenge to the national park ideal until the late 1890s. Park officials ultimately prevailed, but their efforts to create an uninhabited wilderness preserve would have far-reaching consequences that have yet to be resolved.

3

BEFORE THE WILDERNESS

Native Peoples and Yellowstone

> The Great Spirit made these mountains and rivers for us, and all this land. We were told so, and when we go down the river hunting for food we come back here again. We cross over to the other river and we think it is good.
>
> Kam-Ne-But-Sa, or
> Blackfoot (Crow), 1873[1]

ON A LATE SUMMER EVENING IN 1870, at the junction of the Firehole and Gibson Rivers in what is now Yellowstone National Park, some of Montana's leading citizens gathered around a warm campfire for supper and rest. After much hard travel, the first official Yellowstone expedition of "discovery" was drawing to a close, and its members spent much of the evening talking about the many wonders they had seen in the past three weeks. According to popular legend, their conversation soon turned to a discussion of how best to tell the world of their adventures. A few proposed that all should lay claim to several quarter sections of land at the most scenic locales and thus profit from the parade of tourists that was sure to follow. One in their party vehemently disagreed, saying "he did not approve of any of these plans—that there ought to be no private ownership of any portion of that region, but the whole ought to be set apart as a great National Park, and that each one of us ought to make an effort to have this accomplished." All heartily concurred and soon dedicated themselves to the creation of Yellowstone National Park.[2]

This somewhat apocryphal story of several gentlemen around a lonely campfire, debating the future of Yellowstone in the midst of a dark, vast wilderness, has long fascinated generations of park historians and visitors alike.[3] But theirs would not have been the only fire burning in the area that night. No doubt several conversations took place that same evening, in several different languages, that discussed the so-called Washburn expedition's importance to the future of the Yellowstone region. On their first day within the present park boundaries, the explorers came across a large group of Crow hunters and their families, who must have puzzled over the presence of nineteen heavily armed men wandering about the countryside with no apparent purpose or direction in mind. Perhaps they were gold seekers or settlers who would soon place new burdens on the land and its resources. Likewise, groups of Bannock and Shoshone certainly discovered some of the expedition's widely scattered campsites along Yellowstone Lake, and they, too, must have wondered at the size of the Washburn party. And just downriver from the famous campfire discussion, a frequently used trail cut through a clearing where any native traveler could have spied down on the large camp. Of course, Americans and Europeans had long passed through the region, first as trappers and later as transient prospectors, but never had such a large number spent so much time in the Yellowstone basin. Whatever the purpose of this new group, it could not bode well for the peoples of the high mountains.[4]

For their part, the members of the Washburn party were keenly aware of Yellowstone's native inhabitants. Fear of Indian attack led them to request a military escort, and the explorers set up a regular night watch through the first half of their journey.[5] Aside from the hundred or more Crow they warily followed over the course of a week, the party came across abandoned Indian camps on several occasions throughout their trip.[6] They frequently relied on well-used Indian trails and began to relax their guard only once they had spent a number of days blazing their own path through dense stands of timber. Even on the hills south of Lake Yellowstone, where they hacked new trails in a vain search for a lost member of their party, the group discovered an abandoned tepee, a game run used for corralling herds of animals, and stacks of lodge poles left behind for later use.[7]

Ironically, they dismissed these signs as ancient remnants of vanished Indians or, in the case of the large group of Crow they encountered, the aberrant behavior of plains Indians who sought refuge in the mountains. In the face of contrary evidence, the members of the Washburn expedition fell back on common assumptions about "vanishing" Indians and the apparently "pristine" wilderness over which they traveled. As one member of the group would write in a popular literary magazine, "unscientific savage[s]" found little to interest them in the soon-to-be-famous geyser basins. Instead, he supposed that Indians "would give . . . wide birth [to such places], believing them sacred to Satan."[8] The scarcity of game animals and plant foods near the geysers limited any chance that a boisterous party of gentlemen explorers might encounter a group of native hunters, but the idea that "pagan Indians" feared a natural manifestation of Christian hell somehow made better sense. While the contradictory nature of such statements was entirely lost on the members of the Washburn party, they at least jibed with the often repeated tales of a few old trappers about Indian fears of geysers and fumaroles; it mattered little that first word of Yellowstone came from the various tribes of the Rocky Mountain region or that evidence of Indian camps could be found throughout the geyser basins.

Yellowstone's Cultural Landscape Before the Historic Era

In 1870, Yellowstone was not, as one member of the Washburn party described it, a primeval wilderness "never trodden by human footsteps."[9] Rather, it was a landscape that had been shaped by thousands of years of human use and habitation. The earliest archaeological evidence in the park area dates to the end of the last Ice Age, when Paleo-Indian groups moved into the region in the wake of retreating ice floes. Over the course of several millennia, climatic and environmental changes in the Yellowstone area, along with the cumulative effects of long human impact, led to commensurate adaptations in native lifeways. A gradual warming trend in central North America some five thousand years ago, coupled with persistent human predation, led to the extinction of ancient species of bison, mammoth, horse, camel, and other large mammals. In place of big game hunting, a new subsistence culture based on extensive plant gathering and the hunting of smaller game animals developed throughout North America. Characterized by a diet of fish, birds, small seeds, legumes, roots, berries, and game animals like deer, elk, mountain sheep, bison, and antelope, this pattern of subsistence continued in the Yellowstone area until the late nineteenth century.[10]

While small bands of hunters and gatherers made the longest and most persistent use of Yellowstone, larger outlying groups from the eastern and western slopes of the Rocky Mountains also exploited the area on a seasonal basis. Likewise, the future park attracted people from distant locales, and indigenous groups traded with the complex horticultural societies that flourished in the Mississippi and upper Rio Grand valleys more than five centuries ago. Yellowstone possessed one of the richest obsidian deposits in North America, and the use of this highly valued material for blades, tools, and ornaments also made the area important far beyond the Rockies. Samples of Yellowstone obsidian have been found at several Hopewellian sites throughout the central United States, and Obsidian Cliff in the northwest portion of the park is littered with native quarries. Shards of Mississippian pottery that date back several hundred years have also been found in the park, which suggests that local people either borrowed the technology for making pottery or traded for these items over great distances.[11]

Yellowstone had other important but less utilitarian attractions. The many geysers and fumaroles in the area held a particular fascination for both distant travelers and indigenous groups. Hardly a feared aspect of the landscape, these areas contain numerous archaeological sites that indicate prolonged and repeated use.[12] Many native peoples no doubt believed that Yellowstone's thermal features possessed spiritual powers, and contemporary Indians from surrounding reservations continue to attribute special healing properties to the hot mineral waters. Some leave small offerings beside or within the springs, a practice that certainly dates back thousands of years.[13] While cleaning "visitor rubble" from a hot spring in the summer of 1959, for instance, a park ranger retrieved an obsidian arrowhead that was probably placed there as an offering long ago. Besides their spiritual associations, naturally hot water and steam also provided a unique resource for cooking and cleaning and for treating certain materials to make them more pliable. Not surprisingly, the geyser basins attracted curious visitors from faraway places who marveled at the strange sights and sounds and bathed in the waters. Even the Salish people, who generally lived well north of Yellowstone, have a story of indeterminate date that predicted the area's future fame once non-Indians spread the word of its many wonders.[14]

While some native peoples went to the geyser basins to pray, Yellowstone's high mountain peaks often served as important vision quest sites, where individuals from various groups went to fast, pray, meditate, and seek guidance from spiritual helpers. This ritual was a private and solitary affair, but certain locales could attract repeated use over long periods. During the 1872 government survey of Yellowstone, for instance, Ferdinand Hayden and his assistant climbed one of the Grand Teton peaks just south of the park boundary. Upon reaching the summit, the two were disappointed to find they were not the first to climb the mountain. As one of their party later reported, the two men found "a space about sixty feet square, in which there is a curious enclosure, formed with stones, some six feet in height . . . [that] must be several hundred years old."[15] Nothing this large was ever encountered within Yellowstone's present boundaries, but Crow elder Francis Stewart saw evidence of ancient "fasting beds" throughout the park in the 1980s and attested to the special reverence that his people have long held for the entire area.[16]

More than pot shards, obsidian quarries, and fasting beds, the park landscape itself provides the best documentation of native habitation and use of the Yellowstone area. By far the most important tool used to shape pre-Columbian North America was fire, and even within the Yellowstone area purposeful burns may have done more to shape aboriginal landscapes than "natural" or lightning-caused fires. Intentional fire not only prevented the sorts of massive conflagrations that now annually plague western forests but also created and maintained important plant and animal habitats on which native peoples based their lives. Seasonal burns opened up broad savannas favored by ungulates, created "open districts" in the forest that eased travel, and encouraged the growth of valued grasses, shrubs, berries, and tubers.[17] Smaller fires kept favored camping sites clear of underbrush and insect pests and served as an important hunting tool. The members of the Washburn party witnessed a "surround burn" by Crow hunters, who encircled game within a ring of fire and then gradually moved in for the kill.[18]

Besides the use of fire, humans manipulated Yellowstone's environment in the choice of animals they hunted and the measures they took to control or augment certain species. Likewise, the gathering and harvesting of food plants and the rudimentary cultivation or elimination of particular shrubs and tubers created human-dependent species. As in other alpine and subalpine areas in the West, some of these untended plants may have become locally extinct in the past century or diminished to small neglected colonies.[19]

The Undiscovered Peoples of Yellowstone

Many peoples inherited, maintained, and exploited this landscape throughout the historic era, but the native groups with the longest connection to the Yellowstone area at the time of its "discovery" in 1870 were a loose association of bands that anthropologists broadly refer to as the Eastern and Northern Shoshone. These geographic distinctions refer to the locales of major winter camping areas in both the Wind River valley in Wyoming and the headwaters of the Snake and Salmon Rivers in Idaho, but these seasonal groupings did not represent a permanent division. Neither a tribe nor a confederacy of tribelets, the Shoshone of the Rocky Mountain area were a loose assortment of communities tied together by marriage, culture, and language. For the most part, they distinguished themselves by the temporary and long-term ecological adaptations that particular families and bands made. Thus, one group might be known as Agaideka, or Fish Eater, because, for a season or a lifetime, its members depended on fish as their dietary staple. Whole bands of Agaideka could temporarily become Kutsundeka, or Buffalo Eater, if they possessed horses and joined other mounted Shoshone for seasonal buffalo hunts on the plains. Although entire group identities were rarely this fluid, individuals and families often moved throughout Shoshone territory and frequently assumed new subsistence patterns and community affiliations.[20]

Whether directly descended from the Archaic peoples of the Rocky Mountain region or later arrivals, the Shoshone were firmly established within the mountains and along the eastern and western slopes of the Rockies by the end of the fifteenth century.[21] After acquiring horses from their Comanche relatives in the early eighteenth century, some groups developed a new culture of equestrian plains nomadism that would characterize Kutsundeka life until the late nineteenth century. Initially, the equestrian Shoshone rapidly expanded eastward onto the plains and as far north as the Saskatchewan River, but Blackfeet, Siouan, and other Algonquian groups soon acquired horses and firearms and, within one or two generations, drove the Shoshone nomads back toward their original territory.[22] Frequently harassed by more powerful plains groups in the buffalo country and even pursued into the mountains, the equestrian Shoshone ultimately retained much of their old dependence on the resources of the central Rockies.[23] Within a core area that covered much of present-day western Wyoming, southwestern Montana, central and eastern Idaho, and a small portion of northern Utah, these people developed a hybrid culture based on plains buffalo hunting and older patterns of alpine and subalpine hunting, fishing, and gathering.

The seasonal migrations of the equestrian Shoshone closely followed the movements of certain game animals, the annual runs of Pacific salmon, and the ripening or maturation of important food and medicinal plants. From late autumn and through the winter, various bands established camps in the foothills and sheltered valleys of the Rockies, where they supplemented winter food stores by hunting elk, deer, and small game. In spring, some of these bands would come together for buffalo hunts on the western plains south of the Yellowstone River or, at least until 1840, travel to the smaller herds in the eastern Great Basin. Others moved to the headwaters of westward-flowing streams for the spring salmon runs before heading into the mountains to gather plants and hunt bighorn sheep, deer, elk, and buffalo.[24] During the era of the famous fur-trade rendezvous in the 1820s and 1830s, large groups of Shoshone would spend several weeks during midsummer at these intertribal gatherings to trade and barter with trappers and native groups from both sides of the Rockies.[25] Likewise, large intertribal gatherings took place during the late summer camas harvests in eastern Idaho and southwestern Montana. After collecting this important staple food, a defensive alliance of Salish, Nez Perce, Bannock, and Shoshone would travel east across the mountains for several weeks of hunting buffalo in potentially hostile territory. By late autumn, these equestrian Shoshone would again break down into smaller bands and head toward the mountains for another winter of chores, stories, hunting, and trapping.[26]

While bands from other tribes often camped with the Shoshone in the Rocky Mountain area and on the plains, none more closely associated themselves with the Kutsundeka than the Bannock. Related to the Northern Paiute from the high plateau areas of eastern Oregon, the Bannock had thoroughly mixed in with the equestrian Shoshone some time after both had acquired horses. Non-Indian observers often had trouble telling the two apart, but they generally referred to the Bannock as more "aggressive" than the Shoshone. Nevertheless, complaints against "hostile Bannock" raiders and horse thieves were just as frequently applied to bands of equestrian Shoshone as they were to combined groups of Shoshone and Bannock. Because the mixed Shoshone and Bannock consisted exclusively of equestrian communities, this reputation probably derived from their ability to range over a large territory. Consequently, they had more contact with non-Indians and no doubt built up an especially strong resentment against the settlers and emigrants who encroached on their vast hunting and grazing lands. Likewise, in their efforts to compensate for the loss of game and key resources, these groups had more opportunity to raid white communities for food and horses.[27]

Unlike the Kutsundeka, Agaideka, and Bannock, who moved seasonally through Yellowstone and other mountainous areas, one group of Shoshone known as Tukudeka, or Sheep Eater, did not adopt the horse. Instead, they resided almost year-round in high alpine environments. The Tukudeka often hunted, traded, and intermarried with other Shoshone and Bannock, but they remained separate from their more numerous and powerful relatives. Likewise, the Tukudeka of different mountain ranges maintained a separateness from each other and, unlike the equestrian groups who moved from one region to another, tended not to associate with other distant Sheep Eater groups. Perhaps numbering as many as one thousand in the early nineteenth century, most Tukudeka lived in the Sawtooth Mountains of central Idaho, the Bitterroot Mountains of southwestern Montana, the Wind River Mountains of western Wyoming, and the Yellowstone National Park area.[28]

Few outsiders encountered the Tukudeka before the reservation era, and most non-Indian observers described them as impoverished hermits who barely eked out a living in their remote mountain homes. Such perceptions may have derived from encounters with bands of Kutsundeka who, through disease or conflict, had temporarily fallen back on the resources of the mountains and the support of their Tukudeka relatives. On the whole, mountain-dwelling Shoshone may actually have been better off than people who lived in larger communities, and they apparently suffered from fewer wants. By far the most reliable description of the Tukudeka in the prereservation era comes from Osborne Russell, who traded with them in 1834 and 1835. He described a group of some two dozen Sheep Eater he met within the future boundaries of the national park as "neatly clothed in dressed deer and sheepskins of the best quality [who] seemed to be perfectly contented and happy." Rather than horses, they relied on "30 dogs on which they carried their skins, clothing, provisions etc. on their hunting excursions [and] . . . were well armed with bows and arrows pointed with obsidian."[29]

The yearly subsistence cycle of the Sheep Eater, who apparently traveled in small groups of just a few families, centered around the gathering of various plant foods and the pursuit of deer, elk, and bighorn sheep. From late spring to autumn, the large game animals were followed on their migrations to high alpine pastures, where several families might join in a communal hunt. Berries, roots, herbs, nuts, and insects were also gathered, while birds and small mammals were trapped with snares and other devices. As game moved to lower elevations with the coming of winter, the Tukudeka did likewise and spent the coldest months in sheltered glens and valleys.[30] They may have occasionally left the mountains during the annual Pacific salmon runs, where they no doubt traded with other native groups. During their annual treks into the mountains, other Shoshone and Bannock groups also sought out the Tukudeka for trade. The Sheep Eater were especially praised for the quality of their furs, deer and sheep skins, and the powerful bows they manufactured from straightened ram's horn. One of these bows was reportedly worth five buffalo hides and capable of putting a well-shot arrow clear through a charging bull.[31]

Besides the Bannock and various Shoshone groups, Yellowstone was also a seasonal home for the Mountain Crow, who wintered just east and north of the present park boundaries. Also known as the "Main Body," the Mountain Crow were the largest and most powerful of three Crow tribal subdivisions. The other two groups were known as the Kicked-in-the-Belly, who partially separated from the Mountain Crow sometime in the mid-seventeenth century and wintered further east near the Big Horn Mountains, and the River Crow, who ranged north of the Yellowstone River. The forebears of these people may have first encountered this part of the Rocky Mountains more than five hundred years ago, but the Crow probably did not begin to use the area on a regular basis until a century or two later.[32] When early-nineteenth-century European and American traders encountered the Crow on the plains, they immediately referred to them as the

"Rocky Mountain Indians."[33] By this time, Crow use of the future park area mirrored that of the equestrian Shoshone and, as one Crow elder recently put it, the mountains were an important "commissary" where the Indians went to hunt, gather plants, pasture horses, seek assistance from spiritual helpers, take the waters, and look for signs of the First Maker.[34]

Perhaps the clearest expression of the Crow people's reliance on the mountains comes from a speech by a noted chief named Arapooash. As he told a young army lieutenant in the summer of 1830:

> The Crow country is a good country [because] the Great Spirit has put it exactly in the right place. It has snowy mountains and sunny plains . . . and all kinds of . . . good things for every season. When the summer heats scorch the prairies, you can draw up under the mountains, where the air is sweet and cool, the grass fresh, and the bright streams come tumbling out of the snow banks. There you can hunt the elk, the deer and the antelope when their skins are fit for dressing; there you will find plenty of white bears and mountain sheep.[35]

More than a paean to mountain living, however, Arapooash's speech reflected the Crow people's devotion to their entire homeland.[36] Both the plains and the mountains served the tribe well, and seasonal use of one area hardly superseded that of another. Nevertheless, the Mountain Crow differentiated themselves from the River Crow and other tribes by their frequent proximity to the Yellowstone area. Consequently, the future park not only represented a key portion of the tribal homeland but also served as one of the primary "commissaries" of the largest branch of the Crow nation.[37]

The rise of equestrian nomadism on the plains also brought Yellowstone within the orbit of more distant groups in the late eighteenth century. Likewise, the development of new commercial relationships with European and American traders made long-distance travel to Yellowstone more attractive and profitable. Along with the Blackfeet, other equestrian plains tribes occasionally moved into the park area to trap beaver or, perhaps more often, steal the caches of American trappers and then sell them to British traders.[38] Intermontane groups like the Nez Perce, Salish, Kalispel, and Coeur d'Alene also traveled through Yellowstone on their way to the buffalo grounds and trading centers along the Missouri River. For the most part, these distant groups used the future park area only on an irregular and short-term basis.[39] Increased American settlement throughout the Rocky Mountain West made long-distance travel more difficult in the mid-nineteenth century and, except for migratory buffalo hunters from west of the Continental Divide, these groups tended to remain within their shrinking homelands. By the time most western tribes were being forced to settle on often remote reservations, the only groups that continued to use the park area on a regular basis were those with the longest claim on Yellowstone—the Shoshone, Bannock, and Crow.

Throughout the nineteenth century, the peoples who most used the Yellowstone area all suffered from general declines in their populations. Before the first disease epidemics of the eighteenth century, both the Shoshone and Mountain

Crow probably numbered upward of eight thousand. At times, they lost as many as half their people to pathogens for which they had no immunities, but numbers seem to have stabilized fairly well after the earliest exposures to European diseases. Populations failed to recover when regular contact with non-Indians and increased competition for resources with other tribes brought hunger, war, and more sickness. The Northern and Eastern Shoshone, for instance, numbered as few as twenty-five hundred in the early nineteenth century and declined another 20 percent to around two thousand by the 1870s. The Mountain Crow's population generally matched that of the Shoshone, and they experienced a similar overall reduction in numbers. These numbers would hold for both groups through the late 1880s, however, as they managed to supplement reservation supplies with off-reservation resources.[40]

The decline of native populations through the mid-nineteenth century did not necessarily translate into less use of the upper Yellowstone basin. Larger alliances of Shoshone, Bannock, and other western slope tribes moved through Yellowstone in the 1860s as they competed with the increasingly powerful Sioux, Cheyenne, and Arapaho for shrinking herds of buffalo on the plains. More significantly, Yellowstone's elk herds took on even greater significance for some groups of Shoshone and Bannock after a series of unsatisfactory buffalo hunts in the early 1870s.[41] Likewise, the development of small agricultural settlements and mining camps along the western slope of the mountains laid waste to important fishing, hunting, and gathering places. Consequently, Yellowstone attracted heavier use by groups from present-day Idaho and southwestern Montana.[42] For large parties of Crow families, the upper reaches of the Yellowstone River served as an important refuge from the more heavily armed Sioux, who moved with impunity through portions of Crow territory in the late 1860s. Furthermore, less successful hunts on the plains made winter hunting all the more important for the Mountain Crow, who no doubt supplemented dwindling food stores with elk, deer, and small game from the future park area.

The Reservation Era

The United States first recognized native rights to the Yellowstone area in the Fort Laramie Treaty of 1851, when several tribes met with government officials to establish official relations and define their respective territories. While these tribal boundaries would set the geographic parameters of future land cession agreements, treaty commissioners were primarily concerned with expected reprisals against continued overland migration. Consequently, the government sought little more than a peaceful guarantee of American rights to travel across Indian lands.[43] In the effort to stave off possible conflicts between various native groups, as well as guarantee American rights of passage, the commissioners and assembled tribal leaders recognized the eastern third of the future park area as part of the Crow nation. Based on early trappers' encounters with Blackfeet hunters in the same area, the northern section was included in their tribal lands.

The equestrian Shoshone did not participate in the actual negotiations with the United States, but a large party did attend the council proceedings, and the treaty essentially defined the eastern and northern boundaries of their territory. By official default, then, the United States recognized the western portions of the future park as belonging to the Shoshone.[44]

Both the Crow and Shoshone negotiated treaties with the United States in 1868 that allowed Americans to occupy or develop large portions of their territory, including most of the present-day national park. Meeting again at Fort Laramie, the Crow ceded their exclusive rights to all lands south of present-day Montana, east of the Little Big Horn country, and north of the Yellowstone River. The Eastern Shoshone met with treaty commissioners at Fort Bridger and made similar cessions to most of their lands, agreeing to live on a "permanent" reservation in the Wind River valley. A stipulation in the treaty that another reservation be created for the Bannock and Shoshone in their western territories led to a presidential decree in 1869 that established the Fort Hall Reservation in what is now southeastern Idaho.[45] The government also negotiated a treaty with "the [Northern] Shoshones, Bannacks, and Sheepeaters" for a small reservation on the Lemhi River, just a few miles from the place where Lewis and Clark first encountered these people in 1805. Although the Senate failed to ratify this last agreement, its basic precepts were implemented by executive order in 1875.[46]

As noted in the previous chapter, an important component of the two ratified treaties was the stipulation that all tribal members had "the right to hunt on the unoccupied lands of the United States so long as game may be found thereon." Article Four of the unratified "Treaty with the Shoshones, Bannacks, and Sheepeaters" also reserved similar rights, and native peoples from the Crow, Wind River, Fort Hall, and Lemhi reservations continued to use the park area much as they had for generations.[47] Yellowstone also remained home to an estimated two hundred or more Tukudeka.[48] Because they were not present at any of the treaty councils with the government, few believed they had an obligation to make new homes with the Bannock and equestrian Shoshone. Even by the late 1870s, the government agent for the Lemhi Reservation still complained that Sheep Eater had "a great disposition . . . for roaming from point to point in the mountains, making the reservation rather a convenience than a home."[49]

Because the square boundaries of Yellowstone National Park had no direct correspondence to any particular tribal territory, native use of the park area is difficult to document in the prereservation era. Native peoples could not possibly distinguish the lands within the park's future borders from those outside. Consequently, none referred to anything except the geyser basins and deep river canyons in more than general terms. Likewise, few trappers ventured into the park area, and only a handful ever recorded their experiences. The creation of the national park in 1872 put an end to Yellowstone's relative obscurity and soon caused natives and others to focus on the area in new ways. As a result, early park and Bureau of Indian Affairs records contain a wealth of information about Indians in Yellowstone throughout the 1870s and 1880s. Ironically, historians have long argued that native peoples stopped entering the park area sometime before 1872, but the comments of contemporary observers suggest that Indian use of Yellowstone may actually have increased at this time.[50] Throughout the 1870s, for instance, both large and small bands of Crow entered the northern portion of the park along the same trail where the Washburn expedition encountered them in 1870.[51] Likewise, Shoshone, Bannock, and other groups from present-day Idaho, Montana, and Wyoming made frequent trips to the new national park to hunt, gather, take the waters, and visit with relatives from the other side of the mountains.[52]

Because it remained "unoccupied land of the United States," Yellowstone was a game-rich environment that surrounding native groups increasingly exploited. Not surprisingly, government surveyors and early park officials often came across recent signs of purposeful burns, hunting camps, and plant-gathering sites.[53] Increased hunting did not seem to affect Yellowstone's game populations at this time, however, and elk numbers actually increased through the latter half of the nineteenth century. The near extinction of the beaver in the 1840s helped, in that these busy animals competed with elk and other ungulates for many of the same food sources. Indeed, elk populations probably exploded in the early 1840s and

Summer encampment at head of Medicine Lodge Creek, June 1871. Variously identified as Sheep Eater, Shoshone, and Bannock, this group was photographed by W. H. Jackson during the Hayden Survey of the Yellowstone area. Medicine Lodge Creek is seventy miles west of Yellowstone, and this family probably did not travel within the boundaries of the future national park. Nevertheless, the temporary nature of their camp reflects the seasonal mobility of many native peoples in the central Rocky Mountains during the 1870s. (This image reproduced by permission of the Huntington Library, San Marino, California.)

did not begin to level off until the park service implemented aggressive game-control policies in the 1920s. Deer and mountain sheep also remained plentiful and continued to attract native hunters through the late nineteenth century. Only Yellowstone's buffalo herd diminished at this time but mostly because of the success of white hunters, who could sell a bull's head to a taxidermist for as much as $300 in the early 1890s.[54]

Despite a growing awareness that Indians probably outnumbered tourists during the first years of the new national park, officials expressed no opinions about native use of Yellowstone until the late 1870s. The earliest concerns came from the Bureau of Indian Affairs, which sought to curb off-reservation usufruct rights and further reduce tribal land holdings. At a meeting with the Crow in August 1873, however, government treaty commissioners learned that the tribe still considered Yellowstone and the surrounding area as part of their homeland. The commissioners protested that miners had already overrun the mountainous portions of Crow country and, along with the land cessions of 1868, the Indians should rid themselves of the parts of their reservation that bordered the national park. They further asserted that, because the Crow were buffalo hunters, the tribe should have no interest in the mountains. This was news to the Crow, who immediately tried to disabuse the commissioners of this notion. Blackfoot, one of the tribe's principal chiefs, angrily pointed out that "there is much game in the mountains," and he had no desire to forfeit any part of his people's territory.[55]

As Blackfoot made plain, the Crow relied heavily on the Yellowstone region in the 1870s. One of their principal winter camps at this time was located just a few miles downriver from Yellowstone's northern boundary, where Crow hunters stalked the herds of elk that gathered in the lower elevations.[56] While the presence of gold miners along the northeastern boundaries of the park caused no end of concern to the tribe, bands of Crow families still moved through these areas on a regular basis to hunt and collect plants. As their agent complained in 1877, the Crow made such frequent use of the park area that he believed they "differ[ed] but little from all the wild tribes of the mountains, who know nothing of the restraints of civilization."[57]

By the early 1880s, however, Crow use of the northeastern portions of Yellowstone dropped off markedly as the tribe struggled against a number of setbacks. Miners continued to overrun the western end of the reservation, depleting game stocks and destroying important food-gathering sites on both sides of the park boundary line. Likewise, the combined effects of commercial and native hunting in the lower Yellowstone Valley wiped out the few game animals that remained and, with the near extinction of the buffalo, made the Crow almost wholly dependent on agency rations.[58] The death of Blackfoot in 1879 dealt the tribe another serious blow and severely undermined its ability to present a united voice to Washington's demands for more land. Weakened and disunited, the Crow ceded the western fifth of their reservation to the United States in 1880 and, four years later, moved to a new agency almost two hundred miles east of the park. Far from the Rocky Mountains and blocked by expanding ranches and growing settlements, Crow use of the Yellowstone region almost ceased completely.[59]

Like that of the Crow, Shoshone and Bannock use of the park area in the 1870s mirrored patterns established in the prereservation era. The Tukudeka remained in Yellowstone on a more or less permanent basis until 1879, when they were induced to settle on reservations in present-day Idaho and Wyoming. Nevertheless, they spent a good part of subsequent years in their former homes.[60] Likewise, equestrian Shoshone continued to use the park on a seasonal basis, and reservation agents frequently commented on their charges' inability to settle down on the reservation.[61] While the establishment of new mining camps on Yellowstone's northwestern boundary spoiled important hunting and gathering areas in the Gallatin Valley, the rest of the park still harbored a wealth of plant and animal resources.[62] In the southeastern portions, for instance, where bands from the Wind River Reservation would have most frequently entered the park, Captain W. A. Jones followed Indian guides along well-used trails to "perfect" camping areas full of game tracks, berry fields, and excellent horse pastures.[63]

The western and southern approaches to the park remained free of settlements, and bands from the Fort Hall and Lemhi reservations also traveled to the park on a regular basis. As one early observer noted, the absence of nearby white communities allowed "considerable" game populations to thrive in these areas of the park, and native hunters pursued elk, deer, and mountain sheep "to the highest mountain summits [where the game went] to escape the flies and mosquitos" in the summer months.[64] Increased tourism and the growth of new settlements outside the park would soon infringe on native use of Yellowstone, but the area continued to attract people from the Fort Hall, Lemhi, and Wind River reservations until the end of the century.

4

FIRST WILDERNESS

America's Wonderland and Indian Removal from Yellowstone National Park

The Indian difficulty has been cured, the Indians have been forced back on their distant reservations, and the traveler in the park will see or hear no more of them than if he was in the Adirondacks or White Mountains.

George Wingate, 1886[1]

LIKE YELLOWSTONE'S FIRST ADMINISTRATORS, the few tourists who visited the new national park in the early 1870s expressed little or no concern about native peoples. For the most part, they accepted a few old yarns about native fears of the park's thermal features and safely assumed that Indians had long avoided the entire area. Because nearly all those who visited Yellowstone focused their attention on the park's "monumental" features, the geyser basins and the Grand Canyon of the Yellowstone, they had almost no opportunity for encountering Indians in the rest of the park.[2] Likewise, the concentration of tourists in two or three locales made it easy for native peoples to avoid any unwanted contact. Not surprisingly, their preference for staying clear of visitors during the summer tourist season only further confirmed popular assumptions about native fears of the park's strange landscape.

The absence of any concern about Indians reflected the principal motives behind Yellowstone's establishment. The creation of the first national park had less to do with ideas about undisturbed nature than a desire to keep the region's scenic wonders out of the hands of private interests. According to Yellowstone's first historian, Congress acted to create a national park out of a fascination with "the innumerable unique and marvelous wonders of the Yellowstone . . . which could be accomplished only by reserving from settlement the region around them."[3] Few of the park's congressional advocates even hinted that Yellowstone would preserve vanishing landscapes or species, and the Park Act's exhortations "against the wanton destruction of the fish and game" were largely ignored for more than a decade. Consequently, the first efforts to exclude Indians from the park should be viewed within the context of Yellowstone as a national "Wonderland" and against the background of renewed military campaigns to curtail the movements of several western tribes. In the late 1870s, increased visitation and a series of "Indian troubles" soon led to a heightened concern about "marauding savages." Early park officials quickly realized that even the slightest fear of Indian attack could prevent tourists from experiencing all the benefits and enjoyments that Yellowstone had to offer the American people.

In one of the most sensational news stories of 1877, both Yellowstone and the Indian wars captured headlines across the country when the U.S. Army pursued five bands of Nez Perce across the national park. The so-called Nez Perce War involved such personalities as William Tecumseh Sherman and Hin-mah-too-yah-lat-kekht, who would soon be famous among Americans as Chief Joseph. The "war" pitted 2,000 troops against 750 men, women, children, and old people on an 1,100-mile odyssey through present-day Idaho, Wyoming, and Montana. After several violent conflicts with soldiers and settlers, the long, desperate retreat ended just shy of the Canadian border on a cold morning in early October. During their ten-week flight, the Nez Perce spent thirteen days in Yellowstone, where they pastured their horses, raided and accosted a few tourist parties, and searched for a safe passage across the mountains to seek refuge or gain alliance with the Crow.

Contrary to most contemporary accounts, the Nez Perce were neither lost in the park nor surprised and startled by Yellowstone's thermal features. As Yellow Wolf recalled more than fifty years later, the Nez Perce scouts "knew that country well before passing through there in 1877. The hot smoking springs and high-shooting water were nothing new to us."[4] Indians accosting tourists was new, however, and added much to the already sensational appeal of Civil War heroes chasing the elusive Nez Perce through the nation's "Wonderland." Widespread accounts of these events also did much to advertise the new national park, but not in a manner that would bring comfort to those who planned to make Yellowstone into a popular tourist destination.[5]

Much to the alarm and chagrin of park officials, another conflict erupted between the U.S. Army and a few hundred Bannock from the Fort Hall Reservation in the summer of 1878. Following years of near-starvation conditions on the reservation, several bands of Bannock lashed out against the army and local white communities after herds of livestock grazed over the Camas Prairie, one of the most important off-reservation food-gathering areas in central Idaho. Already chafing under increased military supervision at Fort Hall, the Bannock headed east toward Yellowstone, where they were pursued by regular troops and the park superintendent's "party of some 20 well armed, mounted, and equipped, resolute and reliable mountaineer[s]."[6] After raiding horses and frightening a number of tourists, the Bannock were attacked and subdued just east of the park by a platoon of soldiers and Crow scouts under the command of General Nelson A. Miles.[7] Yellowstone's

"Indian troubles" would not go away, however, and the following year park officials braced themselves once again when the so-called Sheep Eater War broke out in central Idaho. Although this last conflict did not cross into the national park, fears that Yellowstone Tukudeka might become involved must have led many to believe that the "nation's playground" had become a yearly battleground.[8]

In many respects, park management in the late 1870s resembled that of a small western military installation. The construction of the first park headquarters in 1879—a heavily fortified blockhouse—wholly reflected concerns about further Indian "depredations." Located on an isolated hill that offered the "best defensive point against Indians," the headquarters building was designed to provide emergency protection for official documents, park personnel, and tourists.[9] Superintendent Philetus Norris, who oversaw the construction of the headquarters and managed the park's defenses during the Bannock War, believed the best course of action lay in convincing "all the surrounding tribes . . . that they can visit the park [only] at the peril of a conflict with . . . the civil and military officers of the government." To these ends, he called on the army to set up a small military post on Yellowstone's western boundary to keep Indians from the Fort Hall and Lemhi reservations from entering the park. These measures were apparently successful, and Norris credited the summer military camp for preventing the Sheep Eater War from spreading into the national park.

Fort Yellowstone, 1879. Frontispiece for *Report upon the Yellowstone National Park to the Secretary of the Interior for the Year 1879*. Chosen for its defensive virtues, the first park headquarters was built during the "Indian troubles" of the late 1870s. Superintendent Philetus Norris described the building as a "first-class . . . block-house 40 by 18 feet, two tall stories high, with . . . an octagon turret or gun-room, 9 feet in diameter and 10 feet high, well loop-holed for rifles, all surmounted by a national flag 53 feet from the ground." (Courtesy of the William Andrews Clark Memorial Library, University of California, Los Angeles.)

Although Norris congratulated himself for having eliminated all "annoyance by Indians during the past season within or near the park" and confidently predicted there would be "no . . . prospect of any during the next," he knew that bands of Crow, Shoshone, and Bannock continued to use the park on a regular basis.[10] Moreover, his discovery of a recently abandoned Sheep Eater camp just a few miles from the new headquarters proved that defensive bulwarks alone could not permanently exclude Indians from the park. The presence of this large camp so close to Mammoth Hot Springs left Norris in "rapt astonishment" and, according to the superintendent, threatened to jeopardize pending leases for the construction of hotels and other tourist amenities.[11] To solve the "problem" of resident Indians in the park, Norris turned for help to the agent at Fort Washakie, who responded by sending a party of Shoshone to escort the Tukudeka to new homes on the Wind River Reservation.[12]

The removal of Yellowstone's last native inhabitants in the fall of 1879 proved a great relief to both Norris and early park concessionaires, but no one could rest easy until "the four Indian tribes owning or frequenting any portion of the park . . . cede[d] and forever abandon[ed] it as well as the adjacent regions."[13] Norris recognized a golden opportunity to achieve this goal in the spring of 1880, when a number of Crow, Shoshone, Bannock, and Sheep Eater traveled to Washington, D.C., to negotiate certain land cessions and railroad rights of way. As superintendent of the park, he believed the government's first order of business should be "fixing the southern border of the Crow Indian Reservation." Because surveyors had set Yellowstone's northern boundary some three miles above the Montana and Wyoming territorial border, a narrow strip of park land also lay within the Crow Reservation. Since first learning of this discrepancy in 1877, Norris advocated a quick resolution of the matter for the better "protection and management of the Yellowstone National Park, especially at its headwaters and main route of access to adjacent settlement."[14]

Traveling at his own expense, Norris arrived too late to influence the final agreements between the government and the Indians from the Fort Hall and Lemhi reservations. Nevertheless, he managed to express his concerns about future use of the national park to the Crow. His warnings mattered little because tribal leaders recognized that new developments were already cutting their people off from the park area. As a growing invasion of gold prospectors established camps along the boundary between the reservation and the park, the Crow found it increasingly difficult to exercise their use rights in the mountains. The superintendent's worries about Indian "ownership" of Yellowstone also proved irrelevant. Suffering from the recent death of Blackfoot, their principal chief and unable to prevent a rush of gold seekers, the resistant but disunited Crow ceded the narrow strip of park land, along with the western fifth of their reservation.[15] In a few brief years, as Crow leader Plenty Coups later recalled, mining towns and agricultural settlements would fill the area, and the Crow were forced to their remaining lands farther east.[16]

Though he could not have wished for a better outcome in the government's dealings with the Crow, Norris always believed his biggest problem lay with the

Shoshone, Bannock, and Sheep Eater. After meeting with the Crow, he traveled directly from Washington to Idaho to personally elicit a "solemn promise from all [the] Indians to abide by the terms of their treaty in Washington, and also that thereafter they would not enter the park."[17] After meeting with the acting agent for both the Fort Hall and Lemhi reservations and then holding council with tribal leaders, Norris felt assured of the Indians' "faithful adherence." The following year, he again renewed these unofficial agreements and happily reported to the secretary of the Interior that the Indians had "sacredly observed" their pledges not to enter the park. Although he never met personally with tribal leaders at the Wind River Reservation, Norris apparently corresponded with the agent at Fort Washakie and felt satisfied that his concerns would be equally respected among the people living there.[18]

These efforts to keep Indians out of the park had as much to do with concerns about tourism as they did a conviction that Yellowstone held no real significance for the surrounding native communities. As he wrote in his first annual report, Norris believed "the isolation of the park . . . and the superstitious awe of the roaring cataracts, sulphur pools, and spouting geysers over the surround-

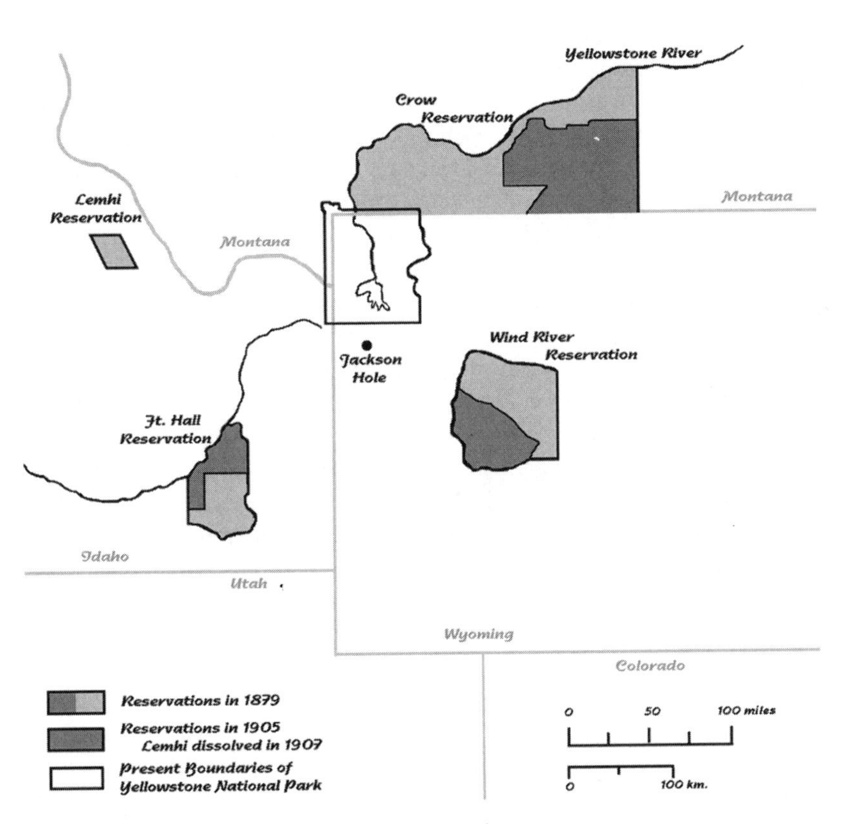

Yellowstone National Park and neighboring reservations.

ing pagan Indians, [caused them to] seldom visit [Yellowstone]."[19] Consequently, his only explanation for the Nez Perce "raid" of 1877 stemmed from their having "acquired sufficient civilization and Christianity to at least overpower their pagan superstitious fear of *earthly* fire-hole basins and brimstone pits." While he viewed the Nez Perce as a partially civilized anomaly, the "few harmless Sheep-eater hermits" were something of a prehistoric anachronism who should willingly abandon their "wilderness haunts" for a better life on a reservation. Ultimately, any Indians who came into the park were ungrateful interlopers, who, instead of appreciating the tireless efforts of reservation agents and Christian missionaries, chose to take advantage of peaceful tourists and the government's unprotected game animals.[20]

Norris apparently had some effect on the Indians with whom he met in 1880 and, as he reported the following year, knew of only one small band of hunters who had entered the park the previous tourist season.[21] Native peoples did not avoid Yellowstone, however, but simply abandoned the more heavily visited areas along the northern stretches of the park. This area had long served as the most important route between the plains and the western slope of the Rocky Mountains, but the near extinction of the bison rendered such travel obsolete. Then again, the loss of the bison herds made seasonal hunts of elk and other game animals in the park area all the more important. Likewise, meager rations and intrusive Americanization programs on the reservation made Yellowstone an attractive haven. Every year, large parties of Shoshone, Bannock, and Sheep Eater left their reservations in Idaho and Wyoming to spend the summer and early fall months along the remote southern and western perimeters of the park—away from tourists and the garrison headquarters but still within long-used areas for hunting, plant harvesting, fasting, and gathering medicinal herbs.[22]

While native use of Yellowstone continued on a seasonal basis, park officials and tourists seemed to have forgotten about Indians altogether by the early 1880s. Superintendent Patrick H. Conger, who succeeded Norris in 1882, made no comment on the neighboring tribes in his first annual report, except to note they were "no longer to be feared."[23] Park visitors shared the superintendent's confidence; as one tourist put it just a few years later, most felt "the Indian difficulty ha[d] been cured."[24] A survey of early guidebooks and visitors' diaries reveals an almost complete lack of interest in Yellowstone's native history and little or no concern about Indian attack. As tourists flipped through the pages of their guidebooks to read about Yellowstone's many natural wonders, they learned nothing of Indians except that fears of the park's thermal wonders had always rendered the area more or less free of the "red man's yell."[25]

The Eden of America

Tourists in Wonderland may not have bothered about anything beyond geysers, canyons, and waterfalls, but a growing number of government officials and influential sport hunting groups began to view Yellowstone as more than a collection of scenic curiosities. In what historian Aubrey Haines has called the beginning of the "Yellowstone war," concern over mismanagement, private leases, and railroad rights of way sparked a series of public debates in the 1880s that led to a redefinition of the purpose and importance of a national park. The troubles began with the Northern Pacific Railroad, whose interest in Yellowstone dated back to its support of the Washburn expedition in 1870. The company's aggressive attempts to control tourist concessions and influence park management had already created powerful enemies in Washington by the late 1870s. When railroad officials lobbied Congress in 1883 to build a line across the northeastern portion of the park, both to transport tourists and to haul gold-bearing ore from mines on the recently ceded lands of the Crow reservation, they created a storm of protest that soon curbed their ambitious plans.[26]

One of the railroad's most powerful opponents and an original supporter of the Yellowstone Park Bill, Senator George Vest, characterized Northern Pacific's efforts as "a gobble by the railway" that would create a virtual monopoly of the tourist trade. Because construction of the line would require Congress to grant several sections of park land to Northern Pacific and permit the cutting of timber and grading of hillsides, Vest believed that allowing a railroad into Yellowstone was a dangerous precedent that "would end in the destruction of the Park."[27] With the support of President Chester Arthur, whom Vest accompanied on a tour of Yellowstone the previous summer, his arguments prevailed; in the censored words of Representative John J. O'Neil, the government refused to allow "one d—d inch of that park [to be] cut off."[28] Northern Pacific would continue its efforts for at least another decade, and subsidiaries of the railroad managed to obtain leases for the construction and management of most concessions, but the proposal to build a line through the park had largely failed by 1886.[29]

The "railroad threat," as it soon came to be known, raised important issues about just what the park's borders were supposed to protect. Of course, no one questioned Yellowstone's significance "as a public park or pleasuring-ground for the benefit and enjoyment of the people," but a new emphasis was placed on the original park act's prohibition "against the wanton destruction of the fish and game."[30] As Secretary of the Interior Lucius Q. C. Lamar noted in the spring of 1886, the chief purpose of a national park had now become "the preservation of the wilderness . . . in as nearly the condition in which we found [it] as possible." For Lamar and his contemporaries, "wilderness" was a fairly simple concept that meant large animal populations and vast stretches of uncut forest. Though trees and game remained plentiful in the park, miners, lumbermen, and hunters might soon prove a greater threat than even the railroad. Indeed, Lamar warned that America's one great forest and game reserve was already under "siege" and needed vigilant protection.[31]

Secretary Lamar's concerns were hardly new, but they had previously taken a backseat to more immediate worries about defense against Indian attacks and improved visitor access to Yellowstone's scenic attractions.[32] As early as 1875, Captain William Ludlow criticized the lack of protection accorded Yellowstone's game animals and argued that only the army could effectively manage the national park.[33] Despite the strength of Ludlow's convictions, it would take General Philip H. Sheridan to make game protection and military management of the park into issues of national concern. In the process, he would bring together a coalition of interests who shared Lamar's belief that wilderness preservation must become the "dominant idea" behind the development and protection of the national park.[34]

After traveling through Yellowstone in the summers of 1881 and 1882, Sheridan complained about the park's inadequacy as a game preserve and offered to provide troops for its protection.[35] Upon returning to Washington in the fall of 1882, the general first appealed to eastern sportsmen and asked them to press the government for greater protection of the park. He quickly garnered the support of several influential senators, who vigorously championed a proposal to bring the park under military management. Sheridan's ideas also received a good deal of coverage from journals such as the *Nation* and *Forest and Stream*, which soon inspired numerous petitions from state and territorial legislatures, sportsmen's groups, and concerned individuals. This widespread support quickly led to the adoption of stronger game rules in 1883 and pushed Congress to authorize the secretary of War to dispatch "the necessary details of troops to prevent . . . [destruction] of the game or objects of curiosity" in the park. Preservationists did not claim success, however, until the military took over complete management of the park three years later.[36]

The First Cavalry to the Rescue

By the time Yellowstone received the protection of the U.S. Army in June 1886, the Shoshone, Bannock, and Sheep Eater once again headed the list of perceived threats to the national park. Defining the value of wilderness in terms of animals and trees led advocates of preservation to view Indians as inherently incapable of appreciating the natural world. Hardly a key component of the wilderness condition, native peoples instead represented the one great flaw in the western landscape. According to the complaints of outdoor enthusiasts in the late nineteenth century, it seemed a wonder that any forests or animals remained in North America since Indians practically based their entire existence on the destruction of wilderness. As early as 1879, sport hunters and settlers complained to the commissioner of Indian Affairs about native hunters who "wantonly destroyed game" throughout the Rocky Mountain region. Even worse, they lit fires "in order to obtain dry fuel for winter use, or to drive the deer to one place where they might be easily killed . . . [and thus] large tracts of valuable timber were burned over."[37]

To most champions of wilderness preservation, the best solution for protecting these areas was an old solution: the use of military force to keep native peoples on their reservations. Such a program would not only preserve wilderness but also fit nicely into ongoing efforts to "civilize" Indians by training them to become self-sufficient agriculturists. These concerns came to Yellowstone in the person of

Captain Moses A. Harris, who served for three years as the first military superintendent of Yellowstone National Park. A hero of the Civil War and veteran of several campaigns against Indians, Harris possessed a great deal of frontier experience that stood him well in his new post. As one historian described him, Harris "was tough as only a frontier cavalryman knew how to be tough [and] . . . he applied all the skill and austere efficiency [to park protection] that he would have devoted to defending a position against an Indian raid."[38]

Within days of his arrival at park headquarters in the summer of 1886, Captain Harris made his first complaint about the one "constant annoyance" that would plague his three-year administration of the national park. Much to his surprise, Harris reported to the secretary of the Interior about a "considerable band of . . . Indians [from the Fort Hall and Lemhi reservations who were] approach[ing] the western boundary of the Park" and quickly realized that native hunters neither feared nor avoided the Yellowstone region. Moreover, as he discovered just a few weeks later, they regularly maintained favored campsites and hunting areas through the use of seasonal burns.[39] The lighting of purposeful fires and hunting within the park seemed to contravene all that Yellowstone now stood for, and Harris took none of his duties more seriously than preventing Indians from entering the area. His efforts would eventually meet with some success but not before he created a public controversy that involved the commissioner of Indian Affairs, the secretary of Interior, several reservation agents, and leading members of the early conservation movement.

Over the course of his tenure as superintendent of the national park, Harris devoted most of his attention to game protection and fire suppression. Though he complained often about white poachers, he believed that one native hunting party "work[ed] more destruction during a summer's hunt than all of the [non-Indian hunters] put together." For Captain Harris, native hunting was an "unmitigated evil" that threatened to undermine the entire purpose of the park. As he confided to the assistant secretary of the Interior, Harris despaired that his "efforts . . . to protect the remnant of the large game of this country and the growing timber in the National Park and adjacent regions" could serve any purpose so long as Yellowstone continued "to afford summer amusement and winter sustenance to a band of savage Indians."[40]

Because Shoshone and Bannock hunting parties traveled in large family groups of twenty-five to a hundred individuals, park officials could at least keep track of their movements. By shadowing the Indians and warning off those who entered the park, these small patrols managed to at least regulate native use of Yellowstone. This was not a satisfactory solution, however, because "constant watch and supervision" required too much effort on the part of Harris's limited staff. Besides, even when Indians could be warned out of the park, they most likely moved in somewhere else. Moreover, abstract boundaries could neither stop game from straying out of the park toward native hunters nor suppress the movement of Indian-caused fires. Although an unlikely increase in park personnel and funding might help, Harris decided early on that only a cooperative effort between park officials and the Indian service could effectively solve Yellowstone's "Indian problem."[41]

When Harris first learned that native hunters were crossing the park's western boundary, he immediately blamed reservation officials for not effectively managing their charges. In subsequent years, he instigated a number of quarrels with the agents at Fort Hall and Lemhi and even appealed directly to the commissioner of Indian Affairs for help.[42] His complaints had some effect, and on May 23, 1888, the commissioner ordered the agents to "let their Indians plainly understand, that the Government would not tolerate hunting, fishing, destruction of timber &c., within [Yellowstone], and to adopt effective measures to keep the Indians . . . away from the Park or its vicinity." Later that same year, after several new complaints from Captain Harris, the commissioner ordered the reservation agents to immediately recall any Indians who were hunting near the park and, if necessary, invoke "the aid of the military to remove them . . . [and to administer] proper measures . . . for their punishment."[43] Agents had a difficult time enforcing the order, especially when it meant that some families would go hungry from lack of rations. The prospect of a trip off-reservation to round up a large band of Indians did not bode well either, and both the Fort Hall and the Lemhi agents repeatedly blamed Harris for exaggerating the supposed threat to the park.[44]

Warnings from their agents did convince the Shoshone and Bannock to camp just outside Yellowstone's boundaries, but small bands continued to enter the park through the fall of 1888. Exasperated that no one could, or would, take adequate measures to keep Indians away from the park and on their reservations, Harris opened up a new barrage of letters to the agents, the commissioner of Indian Affairs, and various officials in the Department of the Interior. As he angrily pointed out in a five-thousand-word report to the Interior secretary, incompetent reservation agents were entirely to blame; their failure to restrict Indian movements not only hurt the park but also prevented their charges from "better . . . cultivating the arts of peace and civilization."[45]

To increase pressure on the Indian service, Harris made his report available to powerful friends in the East, who in turn made the issue of Indian hunting into something of a minor cause célèbre. Some of this support came from men like Theodore Roosevelt and George Bird Grinnell, who had recently established the Boone and Crockett Club to promote an ethic of "manly" restraint among sportsmen. Besides setting proper standards for hunters, the Boone and Crockett Club was committed to "work for the preservation of the large game of this country, and so far as possible to further legislation for that purpose, and to assist in enforcing existing laws."[46] Toward these ends, Grinnell used his position as editor of *Forest and Stream* to support Harris and publicly harangued the Interior Department for not remedying the chronic abuses of "Indian marauders."[47]

Under considerable pressure from several quarters, Indian service efforts to restrict the Bannock and Shoshone to their reservations finally met with some limited success in the early 1890s. Increased allowances for game protection also helped, but subsequent park superintendents still occasionally echoed Harris's earlier complaints against the agents at both Fort Hall and Lemhi.[48] By this time, however, concerns about local white hunters had eclipsed the earlier obsession with native use of Yellowstone. Unlike the Bannock or Shoshone, many non-

Indian commercial hunters specifically targeted the diminishing numbers of bison that remained in the park. As the most powerful symbols of the frontier that Frederick Jackson Turner had recently pronounced closed, the continued survival of the small herd that remained in the park had become a matter of national concern.[49] When park rangers caught a man named Ed Howell with eleven bison carcasses in March 1894, the story proved so sensational that Congress immediately moved to make hunting in the park a federal offense. Within weeks of Howell's capture, a large majority passed a motion from Representative John F. Lacey to give "the rules and regulations made by the Secretary of the Interior for the government of the park, and for protection of the animals, birds, and fish and objects of interest therein," the power of federal law.[50]

While the provisions of the so-called Lacey Act greatly helped park officials curb the actions of local whites, they could do little to restrict native use of Yellowstone so long as Indians exercised treaty rights to hunt off-reservation. However, just as the capture of Ed Howell set in motion a rapid chain of events, another dramatic episode near the park's southern boundary also led to a profound change in federal Indian law. With the support of Yellowstone officials, who knew that native hunters regularly crossed the southern boundaries of the park, a group of Jackson Hole residents banded together in the mid-1890s to put a stop to what they described as "the invasions of Indians . . . whose depredations and outrages . . . are without parallel since the country was settled."[51] In response to similar complaints about the "wanton slaughter of game [animals for their] . . . hides only," the commissioner of Indian Affairs issued an order banning all off-reservation hunting except for the purpose of obtaining meat. As a later investigation discovered, native people steered well clear of white settlements and almost never killed an animal for its hide alone. If anything, the aggrieved parties were the Shoshone and Bannock, who complained of white trophy hunters killing large numbers of animals near their reservations. Nevertheless, Indian movements were closely monitored, and reservation agents increased their supervision of off-reservation hunting.[52]

Most complaints against native hunters came from local guides who could earn a year's livelihood from a few wealthy European or eastern "dudes." Because some of their clients expressed fears about Indians and the success of every native hunter cut into their own pockets, the guides joined with a number of ranchers to make Indian hunting the main issue in the local elections of 1894. Their efforts proved successful, and all the township offices for what is now Jackson, Wyoming, including constable and justice of the peace, were selected because the candidates promised to "take decided steps to . . . keep the Indians out."[53] With the enthusiastic support of Wyoming Governor William A. Richards, local posses under the authority of Constable William Manning resolved to arrest any Indians found hunting in the Jackson Hole region. Although Manning pledged to uphold the new state's recently established game laws, his larger goal was the exclusion of all Indians from the area—particularly those who had made a point to remind him of their rights to "hunt as much as they pleased." Ultimately, he sought the legal nullification of those rights and pursued this goal with deadly conviction. As Manning later described the town's efforts toward these ends, "We knew very well when we started in on this thing that we would bring matters to a head. We knew some one was going to be killed, perhaps on both sides, and we decided the sooner it was done the better, so that we could get the matter to the courts."[54]

After a number of disappointing encounters with bands of Bannock and Shoshone from the Fort Hall reservation, Constable Manning decided that only a large and well-armed posse could effectively check the movements of native hunters. On July 10, 1895, he deputized twenty-six men and then set out to find a large group of Indians he had encountered a few weeks earlier. Three days later, they surprised a camp of twenty-six Bannock; confiscated their property, which included nine tepees, twenty saddles, twenty blankets, seven rifles, one horse, and nine packs of elk meat; and arrested all for violating the game laws of Wyoming. Disarmed, tormented, and forced to march at gunpoint since early dawn, the Bannock grew weary and afraid for their lives when night began to fall. As they were approaching a thick stand of timber, Constable Manning ordered his deputies to load their weapons. The women and children who made up the rear of the procession saw this and cried out in fear, which caused the nine Bannock men in front to bolt for the woods. According to Ben Senowin and other survivors, the posse immediately opened fire, and an old man named Se-we-a-gat was shot in the back four times and killed. Another was injured, and two children were lost. The rest hid among the trees until the following morning and, with the help of other Bannock they encountered on the way, returned to Fort Hall—angry and frightened over the loss of their property and loved ones.[55]

Whatever Manning's original intentions may have been before the shooting, the commissioner of Indian Affairs later described the whole episode as "a premeditated and prearranged plan to kill some Indians and thus stir up sufficient trouble to subsequently get United States troops into the region and ultimately have the Indians shut out from Jackson Hole. The plan was successfully carried out and the desired results obtained."[56] Not surprisingly, the killing of Se-we-a-gat created quite a stir among the residents of Jackson Hole, as well as the Shoshone and Bannock at Fort Hall. Fears of a new "Bannock War" brought five companies of the U.S. Eighth Infantry to the area, and newspapers circulated reports that Bannock warriors were killing settlers and burning homes. Although nothing of the sort actually occurred, Fort Hall agent Thomas B. Teter believed that settlers might attack the two or three hundred Bannock and Shoshone still in the area who refused to come back to the reservation.[57]

After traveling to Jackson Hole, Teter managed to convey to the remaining groups just how perilous their situation had become. Once they returned to the reservation, he made good on a promise to acquire extra rations to compensate for the premature ending of their summer hunt. These efforts did curtail some off-reservation travel, but native leaders remained furious about the possible loss of their treaty rights and, knowing that emergency rations would not last, vowed to cross the mountains again in the fall. By September, Teter had again secured more rations and arranged for the military to escort a group of Bannock to Jackson Hole and recover the articles taken in July. Afterward, several meetings were held with government officials who promised to punish any crimes committed by whites, and tribal leaders decided not to force matters any further. Instead, much as Constable Manning and Governor Richards might have hoped, they agreed to pursue their grievances through legal channels.[58]

To bring a test case before the federal judiciary, the State of Wyoming and the Bureau of Indian Affairs arranged to have someone from Fort Hall arrested for hunting out of season. A Bannock leader named Race Horse agreed to comply with the plan, killed seven elk in the Jackson Hole area, and then submitted himself to arrest for violating the state's game laws. The case was brought before Judge John Riner in the U.S. Circuit Court for the District of Wyoming in early November and, after more than two weeks of hearings, Riner gave his decision on November 21. Based on a strict reading of legal precedent, Riner concluded that treaty rights established in 1868 took precedence over the laws of Wyoming, which was incorporated as a state in 1890. Echoing the words of Chief Justice John Marshall in the landmark case of *Worcester v. Georgia* (1832), Riner agreed with Race Horse's attorneys that a treaty made by the authority of the United States is superior to the constitution and laws of any individual state. Consequently, article four of the 1868 Fort Bridger Treaty still guaranteed to Race Horse the right to hunt on public lands in the State of Wyoming.[59]

When word of Judge Riner's decision reached Fort Hall, the Bannock and Shoshone celebrated and made plans to resume their trips across the southern portions of Yellowstone National Park to Jackson Hole. Fearing a repeat of the previous year's violence, Agent Teter urged the government to negotiate a new agreement with the Fort Hall Indians that would give financial compensation for relinquishing the rights guaranteed them in the 1868 treaty. Before such negotiations could take place, however, the Wyoming attorney general had appealed Riner's decision to the U.S. Supreme Court and managed to gain a hearing by the following March. On May 25, 1896, just one week after rendering its famous "equal but separate" ruling in *Plessy v. Ferguson*, the court reversed Judge Riner's decision in the less noticed but equally important case of *Ward v. Race Horse*. Arguing that the legal precedents that guided the lower court's ruling were "wholly immaterial," the justices believed the 1868 treaty must be viewed in the context of American assumptions at the time—that is, a temporary expedient that government officials expected neither to honor nor to uphold in light of subsequent events.[60]

In the words of Associate Justice Edward Douglas White, who wrote the majority opinion for the court, the Fort Bridger Treaty was negotiated at a time when "the march of advancing civilization foreshadowed the fact that the wilderness which lay on all sides of . . . the reservation was destined to be occupied and settled by the white man, hence interfering with the hitherto untrammeled right of occupancy of the Indian." Even though the Bannock and Shoshone continued to avoid white settlements, "the right to hunt given by the treaty clearly contemplated the disappearance of the conditions therein specified." Moreover, the promise of such rights was not perpetual, as the treaty specifically stated, but temporary, because federal negotiators clearly understood that a state would eventually exist in the area. Because Congress admitted Wyoming into the Union on "an equal footing with the original states in all respects whatever," game laws in Wyoming did not have to recognize treaty rights any more than similar statutes passed in New Jersey or Vermont.[61]

Although Congress may have unilaterally terminated certain provisions of the Fort Bridger Treaty by admitting Wyoming into the Union, the justices concluded that such action was constitutional in that it followed recent precedent. In a remarkable acknowledgment of the intimate link between national parks and native dispossession, the court cited the creation of Yellowstone in 1872, just four years after the treaty, and the subsequent passage of the Lacey Act in 1894 as effective demonstrations of congressional authority to nullify Shoshone and Bannock hunting rights. Justice White would further clarify his views on such matters in *Lone Wolf v. Hitchcock* (1903), but his opinions regarding the Fort Bridger Treaty already demonstrated the fundamental idea behind the decision in that landmark case; namely, treaty rights existed only when Congress chose to honor them. Even unintentionally neglecting to recognize past treaties could be interpreted as an active expression of the government's will. Though nothing in the Yellowstone Park Act mentioned treaty rights, for instance, White believed the creation of the national park should be interpreted as "a clear indication of the sense of Congress on the subject." Consequently, Wyoming could be admitted as a state, and pass laws for its own governance, without considering preexisting treaty rights.[62]

As one legal scholar has described the *Plessy v. Ferguson* decision, the ruling in *Ward v. Race Horse* was based on a "petty rationalization" of contemporary prejudices over constitutional theory. Perhaps for that very reason, the repercussions of the court's ruling have continued for more than a century. In many respects the Indian equivalent of *Plessy*, the decision has not yet been overturned by a case like *Brown v. Board of Education,* and *Ward v. Race Horse* remains the legal basis for restricting all native hunting in the State of Wyoming.[63] The most immediate result of the decision was the authority it gave local officials to arrest any Indians who ventured onto public lands during closed hunting seasons.[64] This not only served to effectively restrict the Shoshone and Bannock to their reservations in Idaho but also became a powerful vindication of the long efforts to exclude all native hunters from Yellowstone. In a remarkable convergence of interests, park officials had supported the original efforts of the Jackson Hole settlers and Justice White had singled out the Yellowstone Park Act as the legal foundation for any efforts to keep Indians off public lands. No superintendent had ever questioned his legal authority to obstruct or prevent native use of the national park, but *Ward v. Race Horse* now obligated various state and federal agencies, including the Bureau of Indian Affairs, to keep native hunters away from Yellowstone and safely confined to their reservations.[65]

Despite the court's ruling, small numbers of Shoshone and Bannock still left their reservations to gather camas in central Idaho or hunt in the Rocky Mountains, and surreptitious use of the Yellowstone area continued for several years. In

Wickiups, Yellowstone National Park. Though many have collapsed in recent years, similar structures are located throughout the park. Early visitors and government explorers often referred to them as "abandoned tepees," but these were actually the permanent frameworks of seasonal lodges. Situated in heavily used areas, these twelve to fifteen foot structures could be covered with foliage, hides, or canvas to provide shelter and warmth. These two particular wickiups are near a creek in the northwest corner of the park and may have been used until the 1880s and 1890s. (Photo courtesy of the National Park Service, Yellowstone National Park Archives.)

December 1897, for instance, army scouts came across a recently abandoned tepee near Cook Peak in the northern part of the park.[66] This was not the last mention of native peoples in Yellowstone, however. Park officials later supported the efforts of a private concessionaire named E. C. Waters to locate a few Indians on Dot Island in Yellowstone Lake, where they would spend the summer months with a small herd of bison. Waters hoped to attract more business for his Yellowstone Lake Boat Company by creating the sort of "aboriginal exhibit" that had fascinated visitors at the World's Columbian Exposition in Chicago just a few years previously. Park authorities and the secretary of the Interior made only one stipulation: Waters needed to use Crow Indians instead of Shoshone or Bannock.[67] Though he soon moved some animals onto the island, he apparently had no luck convincing any Crow to camp in the middle of Yellowstone Lake. With the failure of this tourist display, Yellowstone finally became the non-Indian wilderness it was always intended to be—both in fact and in the historical imaginings of tourists and park officials.

As the National Park Service proudly proclaims, Yellowstone provided the archetype for later preservation efforts and continues to serve as a model for late-twentieth-century conceptions of wilderness. This chapter demonstrates that

Yellowstone also provides the first example of removing a native population in order to "preserve" nature. As an empty, seemingly untouched landscape, locked away and undiscovered for centuries, Yellowstone represents a perfect Eden, a virtual manifestation of God's original design for America.[68] This conception of wilderness preceded the creation of the first national park by a number of years and proved so powerful that early preservationists either dismissed or ignored any evidence of native use and habitation. And later, when park officials did take notice of Indians, they viewed native hunters as a dangerous and unnatural threat to Yellowstone's fragile environment—even when government surveys demonstrated that populations of most game animals in the park continued to increase through the late nineteenth and early twentieth centuries.[69] These ideas shaped park policy for three decades, until Yellowstone had indeed become a place that native people neither used nor occupied.

Persistent native use of the park area made these fictions difficult to maintain, however, and Yellowstone's early history demonstrates that the creation of uninhabited wilderness required a great deal of effort. Laws and policies may have reified a certain conception of wilderness, and thus provided a model for other preservationist efforts, but in doing so they perpetuated a difficult set of unresolved problems. If anything, Yellowstone served as a prelude for later conflicts between government officials and native groups at other national parks. These new struggles would both strengthen and reshape old ideas, but they did not make the task of creating and preserving uninhabited wilderness any easier. Yellowstone remains a contested place, and native peoples have challenged their exclusion from the park for more than a century, but the issues they raised have played out most powerfully in the history of Glacier National Park. Since that park's establishment in 1910, Blackfeet efforts to exercise certain usufruct rights in the Glacier area have continued to present one of the strongest challenges to the American preservationist ideal.

5

BACKBONE OF THE WORLD

The Blackfeet and the Glacier National Park Area

> [After Old Man created the world and taught the first people how to live,] he traveled northward and came to a fine hill. He climbed to the top of it, and there sat down to rest. He looked over the country below him, and it pleased him. Before him the hill was steep, and he said to himself, "Well, this is a fine place for sliding; I will have some fun," and he began to slide down the hill. The marks where he slid down are to be seen yet, and the place is known to all the people as the "Old Man's Sliding Ground." . . . In later times once, [Old Man] said, "Here I will mark you off a piece of ground," and he did so. Then he said: "There is your land, and it is full of all kinds of animals, and many things grow in this land. Let no other people come into it."
>
> "Blackfoot Genesis," as recorded by George Bird Grinnell, 1892[1]

THE CREATION AND EARLY MANAGEMENT OF Glacier National Park reflected the maturation of American ideas about wilderness as scenic playground, national symbol, and sacred remnant of God's original handiwork. The first decades of the twentieth century also marked a heightened interest in the "vanishing Indian," and the Blackfeet, whose reservation borders the park on the east, became an important feature in early tourist promotions. While the presence of Indian dancers in front of the park's grand hotels tantalized visitors who had come to "meet noble [Indians in] . . . their native home," park officials vigorously enforced a series of programs that excluded the Blackfeet from the rest of the park.[2] The importance of Indians to the Glacier tourist experience seems odd when juxtaposed with policies that actively tried to prevent native use of the park's backcountry, but this apparent irony holds an internal consistency when viewed in terms of early-twentieth-century ideas about Indians and wilderness. As "past-tense" Indians, those Blackfeet men and women who entertained tourists were presented as living museum specimens who no longer used the

Glacier wilderness—if, in fact, they ever did. Those Indians who continued to use the park illegally were simply un-American in their lack of appreciation for the national park and almost barbaric in their unwillingness to let go of traditional practices.

The eastern half of Glacier National Park was once part of the Blackfeet reservation, and the tribe has long maintained that an 1895 agreement with the United States permanently reserved certain usufruct rights within the park area. The National Park Service, however, has repeatedly argued that the Glacier National Park Act of 1910 extinguished all Blackfeet claims to the mountains on the western boundary of their reservation. The impasse hinges on two very different conceptions of Glacier's landscape, both of which reflect deeply held ideas about national identity and cultural persistence. For many Blackfeet, their "illegal" use of the Glacier backcountry preserved a connection to places and items that had been important to the tribe since time out of memory. Indeed, surreptitious use of certain plants, animals, and religious sites within the park helped preserve a wealth of knowledge that would otherwise have been lost forever. Furthermore, Blackfeet use of park lands and resources illustrated de facto proof of the tribe's political sovereignty as recognized in treaties with the United States. For Americans, however, Glacier was one of the nation's most spectacular "crown jewels," and Blackfeet use of park lands threatened to tarnish its luster.

Non-Indians rarely took Blackfeet claims to Glacier seriously and argued instead that efforts to exercise treaty rights or receive compensation for their loss represented little more than selfish opportunism on the part of tribal leaders. Besides, claims that the mountains in Glacier had any traditional importance to the Blackfeet wholly contradicted popular conceptions of proud Indian warriors aimlessly roaming about the flat plains in search of enemies or buffalo. More important, Indian "poaching" and "trespassing" on park lands demonstrated a pronounced disdain for the ideals of wilderness preservation. Although Blackfeet use of park lands certainly contradicted popular stereotypes about Indians and undermined the ideal of a pristine, uninhabited wilderness, the tribe's legal claim on the park proved a much more serious matter for park officials. Ultimately, it called into question the symbolic potency of all national parks.

The term Blackfeet can be confusing and deserves some clarification. It specifically refers to the Pikuni or Piegan Indians residing on the Blackfeet reservation in northern Montana who officially refer to themselves as the Blackfeet Nation. The Pikuni are historically and culturally affiliated with the Siksika (Blackfoot) and Kaina (Blood), and together they comprise the Nitsitapii. More commonly known as the Blackfoot Confederacy, the Nitsitapii is made up of three bands that are divided between Canada and the United States. The Kaina, Siksika, and North Piegan live on three reserves in Alberta, while the largest group within the confederacy, the South Piegan or Blackfeet, live on their reservation in Montana. The Glacier area is important to all these groups, but the Blackfeet have the strongest connection.[3]

Although the Blackfeet maintain an especially deep attachment to the mountains that border their reservation, they are not the only native group with a strong connection to the region. The confederated Salish-Kootenai tribes of western Montana have long frequented the mountains now contained within the national park. Other groups that used the Glacier area in historic times included the Kalispel (Pend d'Orielle) and bands of Crow, Atsina (Gros Ventre), Nakota (Stoney), Cree, and Assiniboine. Along with many Blackfeet, some Cree who live on the Blackfeet reservation, as well as people from the Salish-Kootenai reservation, continue to use park lands to this day. Unlike these other groups, however, the Blackfeet were featured extensively in park promotions from the 1910s through the 1930s, and only the Blackfeet had recognized treaty rights to Glacier National Park at the time of its establishment in 1910.[4]

Backbone of the World

For centuries the Blackfeet have regarded the Glacier area as part of Mistakis, the Backbone of the World. The mountains marked the outer edges of tribe's territory, but they did not stand at the periphery of the Blackfeet world. Within the mountains lived powerful spirits such as Wind Maker, Cold Maker, Thunder, and Snow Shrinker (Chinook winds). One of the most important characters in Blackfeet mythology, a trickster called Napi or Old Man, created the mountains, rivers, prairies, hills, forests, and all the animals of the Blackfeet country. He then created the Blackfeet and taught them how to hunt and gather plants in the mountains and on the plains. Many often-told stories detail Napi's adventures in Mistakis, and he is attributed with the origination of many of the tribe's most important ceremonies, spiritual practices, and everyday customs. Though beneficent, Napi could also withhold things from the Blackfeet, as he did when he punished them by hiding all the bison far up Cut Bank Canyon, in what is now the national park. When Napi left his people in the Long Ago Time, he disappeared into the Rockies, and ever since the Blackfeet have looked upon the Glacier region as a land special to their Creator.[5]

Besides Napi, other important teachers and forces resided in the mountains, including grizzly bears, eagles, ravens, and beavers. The Glacier area was the source of the Beaver Pipe Bundle, one of the most venerated and powerful spiritual possessions of the tribe. Given to the Blackfeet by the Beaver People in the Long Ago Time, the bundle contained numerous sacred items associated with the Backbone of the World and served as the focus of the tribe's most important religious ceremonies. The Beaver People were also the source of the Sacred Tobacco, which they gave to the Blackfeet at one of the Big Inside Lakes within present-day Glacier National Park. For that reason, elders would often plant tobacco gardens near St. Mary, Waterton, and Two Medicine Lakes.[6] Aside from beavers, other powerful Underwater People lived in Mistakis and provided many of the most important staples of Blackfeet life. A story about the first horses, for instance, tells how they were given to a Blackfeet boy by a family who lived in a tepee at the bottom of Upper St. Mary Lake.[7]

The Backbone of the World was also the abode of the Thunder Bird, who gave the Blackfeet their first Medicine Pipe and was said to reside most often in a cave on the side of Chief Mountain. This mountain, which stands at the border of the reservation and the national park, is one of the most distinct and spiritually charged land features within the Blackfeet universe and continues to be an important locale for vision quests and traditional ceremonies.[8] The mountains also provided other sacred materials for the Blackfeet, and at least until the early twentieth century some ceremonial leaders would collect certain mineral and plant dyes to make the face paints used during the Sun Dance and other celebrations.[9]

Recent archaeological studies suggest that Blackfeet use of the Glacier area stretches back over a thousand years.[10] In the centuries before the arrival of the horse, the Blackfeet probably based much of their livelihood on the resources of the mountains and eastern foothills now within the national park. Numerous *piskuns,* or buffalo jumps, are located along river bluffs in the broken country just east of the mountains, where whole communities would combine their efforts to drive part of a herd over the edge of a cliff. Though an important part of pre-equestrian life, bison hunting could never provide the Blackfeet with all their needs. Consequently, seasonal movements in the days before the horse kept close to the mountains to exploit the tremendous variety of plants and animals there at different times of the year. A myriad of food and medicinal plants could be found only in the foothills and mountains, and bighorn sheep, mountain goats, elk, varieties of deer, and smaller animals of the Glacier region also provided the Blackfeet with important meat and hide resources.[11]

The introduction of the horse into Blackfeet culture, which probably came from the Northern Shoshone in the first half of the eighteenth century, and entrance into the European fur trade shortly afterward initiated profound changes that fostered a more mobile and far-ranging annual cycle.[12] Nevertheless, the pre-reservation era Blackfeet still followed an annual round that reflected patterns established by previous generations. The yearly cycle began in early spring, as individual bands left their winter camps in the broken country to the east of the mountains for an intensive season of hunting and plant collecting. Women and youngsters went to the foothills to dig roots and collect other plant foods, while small bands of hunters headed east for bison. Groups of men and women would also head to the mountains to cut and prepare lodge poles, while certain elders planted the sacred tobacco. Plant gathering in the mountains continued through the summer months until the annual Sun Dance, when disparate groups would convene for several weeks at a predetermined location. This annual event would often take place near the Sweet Grass Hills, a favored hunting area about ninety miles east of the Continental Divide.

At the conclusion of the ceremony, which lasted one to two weeks, the various bands would disperse again to hunt bison and collect ripening fruits. Some followed the large herds eastward, while others moved into the mountain valleys and upland areas along the eastern slope of the Rockies to hunt elk, deer, bighorn sheep, and mountain goats; collect prepared lodge poles; and gather berries through the early autumn months. As fall came on, the bison herds moved westward and northward to their wintering grounds, and the Blackfeet as a whole would reassemble into larger groups for communal hunts. Some of this hunting undoubtedly took place within the eastern valleys of the future national park because numerous bison skeletons could be found there as late as the 1890s. The annual cycle of hunting and harvesting would end with the establishment of winter camps in low-lying, heavily wooded river valleys near the mountains—screened from severe north winds by the tall peaks and close to the winter foraging areas of elk and deer.[13]

Though largely dependent on the bison for trade and sustenance, Blackfeet culture and economy were far more diverse than nineteenth-century traders and government officials realized. As men, they had few opportunities to observe the role of women in Blackfeet society. Most of their dealings with the tribe occurred during formal gatherings of all-male councils. Because their contact with the Blackfeet was generally limited to the summer months, when most bands were hunting bison on the plains, early commentators had very little sense of how the Blackfeet spent most of each year near the mountains. Likewise, these early observers tended to draw on their own martial backgrounds and focused their attention on the wide-ranging hunting and warring campaigns that took place in late summer. Not surprisingly, they largely overlooked those aspects of Blackfeet life that did not directly relate to bison hunts and war campaigns.

Although these men provided the base material for later stereotypes about Plains Indian society, they also grossly misunderstood the importance of women in Blackfeet culture. Aside from their role in the preparation for large bison hunts and the rendering of hides, meat, and other products, Blackfeet women were largely responsible for exploiting the nonanimal resources of the mountains and foothills. This geographic split in men's and women's roles, especially the primary importance that the Blackfeet placed on women's contributions to the livelihood of the tribe, is reflected in a version of the Blackfeet creation story. As Tail Feathers Coming over the Hill told it in 1916, men and women once lived separately, and "the camps of the two sexes were far apart: the women were living . . . at the foot of the mountains, in Cutbank Valley, and the men were way down on the Two Medicine River." When Old Man decided to bring them together, he told the men to go to the mountains and join the women.[14] While this story explains how men and women came to live together, it also reflects the glad reunions that many couples and families must have experienced at the end of the summer hunting and gathering season.

The Blackfeet entered into their first treaty with the United States in 1855, but these official relations did little to change long-established patterns. By the late 1880s, however, disease, war, famine, and the near extinction of the bison had reduced the Blackfeet to some two thousand individuals. At the same time, a series of land cessions dramatically eroded the tribe's land base and forced the Blackfeet to establish semipermanent communities along the foothills of the northern Rockies. The mismanagement and corruption of government officials made this adjustment even more difficult, but the Blackfeet persisted and made the reservation their tribal home.[15] As Chief White Calf described it in the late 1880s, the reservation had become the tribal "body." For White Calf and the Blackfeet, a person's body represented intentions and character, as well as possessions and limitations. Thus, to call the reservation his people's "body" was to regard it as the place that belonged to and defined the Blackfeet, and the place to which they belonged as well.[16]

Making up the western portion of the reservation, the mountains remained an important part of Blackfeet life, and most families located themselves in small communities near the foothills. During the painful adjustment to reservation life, the Blackfeet developed a new dependence on the Glacier area that allowed them to maintain older traditions and ameliorate the loss of others. Though no longer able to hunt bison, for instance, young men could still prove their worth as they sought out deer, elk, sheep, and small game in the mountains. Women supplemented meager government rations with the traditional foods and herbs they gathered in the alpine environments, and healers collected and tended medicinal plants. In the midst of pervasive Americanization programs, the Blackfeet also turned to the shelter of the Backbone of the World to hold prohibited ceremonies. Likewise, young traditionalists maintained their connections with the past by fasting in the same remote locales as their forebears.[17] The mountains also provided the resources that made the incorporation of new skills and livelihoods possible. Along with firewood and lodge poles for tepees, high-elevation forests became an important source of timber for the construction of cabins, fences, and corrals. The foothills sheltered some of the best pasturage for new herds of livestock, and the Indian Service tapped into lakes and streams to create a series of irrigation projects.[18]

George Bird Grinnell and the Crown of the Continent

As the Blackfeet struggled with life on the reservation, late-nineteenth-century wilderness enthusiasts came to appreciate the mountains by a very different route. George Bird Grinnell perhaps most fully represented his contemporaries' ideas about Indians and wilderness, and it is through him that preservationist ideals came into conflict with Blackfeet use of the Glacier area. Early in his career, Grinnell had served as a government scientist on Lieutenant Colonel George A. Custer's reconnaissance of the Black Hills in 1874 and the U.S. Army's exploration of Yellowstone National Park in 1875. Despite his early connection to such famous people and places, Grinnell would become even better known for his leading role in the effort to establish Glacier National Park and for his interest in the Blackfeet. As editor of *Forest and Stream* and the author of numerous books on Indians and the West, Grinnell was a leading voice for the preservation of wilderness landscapes and a respected advocate of Indian policy reform. Because his family background gave him strong connections among Washington's powerful elite, his views often found a receptive audience in the capitol and frequently shaped federal policy as well.[19]

On his first visit to northern Montana in 1885, Grinnell became instantly enamored of the mountains within the Blackfeet reservation. For the next several

Blackfeet Sundance encampment at Pray Lake, ca. 1914. Photographed by R. E. "Ted" Marble, this particular camp may have been staged for a motion picture about the Sundance. It is located at a frequently and long-used camp site in the Two Medicine area of Glacier National Park. Both before and after the establishment of the national park in 1910, Two Medicine proved an important area for hunting, plant gathering, and vision questing. The area is also associated with the holding of important Sundance ceremonies in the early reservation era. (Courtesy of the Glacier National Park Archives.)

years, he returned to hunt and explore what he called the last remaining "wild and unknown portion" of the United States, and he published several articles about his adventures. In search of untrodden, pristine landscapes, Grinnell relied on Blackfeet guides and followed countless Indian trails to discover areas that he described as "absolutely virgin ground . . . with no sign of previous passage."[20] The irony of such statements was lost on Grinnell, however, and his enthusiasm for what he called the "Crown of the Continent" wholly supplanted native concerns about the area. During a trip to the mountains in the late summer of 1891, Grinnell resolved "to start a movement to buy the [area] . . . from the Piegan [I]ndians at a fair valuation and turn it into a national reservation or park." He assumed that "the Great Northern R[ailroad] would probably back the scheme and . . . all the Indians would like it."[21] Grinnell believed that, as "primitive" people just setting out on the road to "civilization," the Blackfeet would only be too happy to sell an area of little importance to their future social "evolution." Although he correctly assumed the support of Great Northern, then completing its rail line through the southern portion of the reservation, Grinnell could not have been more wrong about the importance of this region to the Blackfeet.

Initially, Grinnell came to the Rockies to partake of the outdoor life, but he also developed a great interest in his Blackfeet hosts. During each of his visits to northern Montana, he recorded the stories and memories of tribal elders who had come to social prominence during the "Buffalo Days." In addition, he exposed the corruption of agency personnel and successfully lobbied Washington to improve conditions on the reservation. The miserable state of Indians on their reservation was wholly divorced from concerns about his "beloved mountains," however, and Grinnell never linked the fate of the Blackfeet with his plans to preserve the Glacier area.

Like many Americans, Grinnell lamented the rapid exploitation of the western wilderness as he bemoaned the destruction of native societies, and his efforts to preserve some remnant of each epitomized late-nineteenth-century thinking about wilderness as uninhabited and Indian culture as vanishing. Not surprisingly, his efforts to preserve Blackfeet culture and some portion of the tribe's homeland took very separate courses: the Blackfeet would live on in books and museum collections, but the mountain wilderness would persist within the boundaries of a national park. Grinnell's concerns not only reflected a sharp distinction between Indians and wilderness but also grew out of popular ideas about the place of each within an increasingly industrialized and urbanized America. Consequently, he encouraged his countrymen to "uncivilize" themselves a bit and return to the mountains on a regular basis but admonished his Blackfeet friends to become "civilized" and enter the mainstream of American society.[22]

In the early 1890s, rumors of mineral wealth in Grinnell's favorite stomping grounds threatened to destroy his plans to convert the region into a nature preserve. The influx of fortune hunters wreaked havoc on the reservation and occupied most of the tribal police's time in a losing effort to evict trespassers and curb their abuses. Reservation officials, who had a vested interest in keeping order on Blackfeet lands and hoped to stake claims of their own in the mountains, made common cause with mining interests and successfully petitioned the government for a cession of the western part of the reservation. Washington acted swiftly and, in the summer of 1895, the commissioner of Indian Affairs asked Grinnell to help negotiate a land cession agreement with the Blackfeet.[23] Although Grinnell hated to see the mountains overrun with miners, he was confident the area possessed no great mineral wealth. Instead, he viewed the cession of Indian lands as an important first step in the creation of a great national park. Even before meeting with the Blackfeet, Grinnell sought to interest others in his plans for the proposed cession of land. Most significantly, he interrupted his journey from New York to northern Montana in St. Paul, where he personally lobbied officials of the Great Northern Railroad to support his efforts to create a national park.[24]

The Ceded Strip

Tribal leaders respected Grinnell and even requested that he be one of the government commissioners, but negotiations were contentious and at one point broke down altogether. As Little Dog reminded the commissioners at the start of the proceedings, "The Indians did not ask the government to come and buy their land," and most still felt bitter about previous treaty violations.[25] Although the Blackfeet had no great desire to sell any more land, they were painfully aware that

Place without people. Charles M. Russell, *When the Land Belonged to God*, 1914. Usually populated with cowboys and Indians, Russell's most famous paintings portrayed a romantic vision of nineteenth-century life in the American West. In this remarkable painting of bison fording a river, humans are entirely absent and, as Russell's title declares, the land belongs entirely to God. Such a vision of pristine, uninhabited wilderness had already informed George Bird Grinnell's efforts to create a national park on Blackfeet lands. (Courtesy of the Montana Historical Society.)

funds from an 1887 land cession had nearly run out without their realizing any appreciable gains. Consequently, some viewed the sale of the mountainous portion of their reservation as an opportunity to gain economic self-sufficiency and offset some of "the many things in which the Great Father has cheated us."[26] Toward these ends, tribal leaders agreed to sell part of the land desired by the commissioners, but at nearly triple the government's asking price. When Grinnell and the other commissioners balked at this offer and then chided the Blackfeet for their "folly," talks broke off acrimoniously.

The Blackfeet were negotiating from a weak position, however, and failure to reach an agreement with the government might have proven catastrophic. Nothing could prevent miners from invading the reservation, and the tribe might eventually have to give up the land without any compensation. Without sufficient funds or government support, the specter of another starvation winter like the one that claimed nearly a quarter of the Blackfeet just ten years earlier must have seemed imminent. Out of this sense of desperation, tribal negotiators met unofficially with their agent and, after suffering through a late-night session of strong-arm tactics, agreed to meet one more time and sell the land requested by the commissioners for $1.5 million.[27]

Despite the almost unavoidable necessity of once again selling land to the United States, a number of important Blackfeet leaders refused to come to any agreement with the commissioners. Moreover, a few dozen eligible participants chose to stay away from the most important tribal gathering of the year and hunted in the mountains instead.[28] Ultimately, such resistance had a profound effect on the final agreement between the commissioners and the Indians. As he accepted the agreement on behalf of the Blackfeet, Chief White Calf could not help but reflect on the mortal blow the entire process had dealt them. Once again conjuring up the image of the reservation as the tribal body, he stated: "Chief Mountain is my head. Now my head is cut off. The mountains have been my last refuge."[29] To alleviate some of the damage, however, and to help ensure that the Blackfeet would continue as a people, he stipulated that the United States must guarantee certain usufruct rights in the ceded area. Various band leaders were quick to second this addition to the agreement, and it received enthusiastic support from all the assembled Indians.

The commissioners acquiesced, and all tribal members retained "the right to go upon any portion of the lands . . . to cut and remove timber for agency and . . . personal uses . . . [and] to hunt upon said lands and to fish in the streams thereof, so long as . . . they remain public lands of the United States."[30] While the term "public lands" may have seemed a vague legalism to some Blackfeet leaders when the commissioners read the final agreement aloud, they could reasonably expect to retain their rights in perpetuity; the area had no agricultural value and, as Grinnell apparently reassured them, it would never support long-term mining.[31] The Blackfeet may also have proved amenable to these conditions because they apparently had a very different understanding of the lands in question. As White Calf described the agreement, his people ceded only those parts of the mountains that lay above timber line. Thus, the Blackfeet reserved all "the timber and grazing lands" and retained the right to hunt at higher elevations. In either case, both the government and the Blackfeet agreed that tribal members could use the entire area in much the same way as before.[32]

Aside from the "public lands" clause, which is absent from any record of the actual negotiations, the commissioners also inserted another provision that would have far-reaching consequences. No doubt aware of the dramatic events that had recently taken place in Jackson Hole, they further qualified Blackfeet rights in the area by making them subject to "the game and fish laws of the State of Montana."[33] As historian Louis Warren has noted, it is very unlikely that tribal leaders would have accepted this particular restriction had they known the state had banned elk hunting in 1893.[34] Indeed, there is much in the written document that tribal leaders probably never discussed with the commissioners. Blackfeet oral history even suggests that the land cession agreement may have been little more than a short-term lease of mineral rights. If mining operations proved successful, the Blackfeet would lose access to a good portion of the so-called ceded strip. Otherwise, they believed the agreement placed no restrictions on their ongoing use of the area. In later years, some Blackfeet would describe the 1895 agreement as the selling of "rocks only," while others recalled that tribal leaders had negotiated for a recession of all lands after fifty years, but no one expressed any thoughts about the final document until the 1910s.[35]

The Senate ratified the land cession agreement within nine months, but the government could not fully survey the ceded area and open it to mining claimants until April 1898.[36] Still, prospectors trespassed on the reservation in growing numbers, and both Grinnell and his friends in the U.S. Geological Survey worried about the effects of miners' fires on the forests and watersheds of the Glacier area. Consequently, Grinnell worked to have the ceded lands included within a proposed forest reserve that Gifford Pinchot, John Muir, Charles Sargent, and others were then surveying on the western side of the Continental Divide.[37] His efforts succeeded, and on February 22, 1897, President Grover Cleveland signed a proclamation establishing the Lewis and Clark Forest Reserve at the headwaters of the Missouri, Columbia, and Saskatchewan rivers. In doing so, the president made special note that Blackfeet "rights and privileges . . . respecting that portion of their reservation relinquished to the United States . . . shall be in no way infringed or modified."[38]

Native leaders had already become aware of the proposed reserve in the summer of 1896, when a young Walter McClintock, who worked as the photographer for Pinchot's surveying party, first met the Blackfeet and began a long career of ethnological study among the tribe. As McClintock explained to Chief Mad Wolf, he "had come from the Great Father . . . for the purpose of protecting the forests of their country, that they might be preserved for future generations."[39] McClintock did not comment on how the Blackfeet received his report, but none of the headmen with whom he met ever expressed concern over the creation of the national forest reserve. If anything, such a reserve would only have curtailed some of the damage wrought by prospectors and thus further preserved the resources on which many Blackfeet families depended.

As Grinnell predicted, the region held no great mineral wealth, and the ephemeral boom busted in a few short years. By the turn of the century, he had brought together a coalition of wilderness enthusiasts, senators, congressmen, and railroad magnates in a campaign to convert a portion of the Lewis and Clark Forest Reserve into a national park.[40] After ten years of hard lobbying, his efforts proved successful on May 11, 1910, when the Glacier National Park Bill became law. Unlike the proclamations that established the forest reserve, however, the Glacier National Park Act made no mention of the rights reserved to the Blackfeet in the 1895 land-cession agreement. Nevertheless, there is no evidence that any Indians opposed the creation of the national park; a few strongly supported the new park, particularly those members of the tribal business council who saw it as a potential boon to the reservation's economy.[41] Besides, the park act's stipulation that Glacier would be removed from "settlement, occupancy, or disposal . . . and set apart as a public park" seemed to be a solid guarantee of the ceded area's "public land" status and thus ensured that Blackfeet rights would remain intact.[42]

6

CROWNING THE CONTINENT

The American Wilderness Ideal and Blackfeet Exclusion from Glacier National Park

> Here is where God sat when he made *America*. . . . A half-hour's stumble brings the sightseer to . . . a scene of beauty beyond all words and retrospect. The setting is all of another age—before man took dominion over the earth.
>
> Tom Dillon (1912)[1]

IN THE FIRST YEARS AFTER THE creation of Glacier National Park, neither tourists nor park officials seemed to trouble themselves about Blackfeet rights in the park. Indeed, early advertisements for Glacier closely identified the tribe with the new park. As the most influential force in developing and promoting Glacier, the Great Northern Railway used its vast public relations machinery to plant photographs and stories about the Blackfeet in magazines and newspapers throughout the country and produced countless illustrated brochures featuring Indians amid spectacular alpine scenery. As if that were not enough, Great Northern President Louis Hill arranged for groups of Blackfeet to travel to New York, Chicago, San Francisco, Minneapolis, and other cities to set up tepee camps on the roofs of downtown buildings. Always well covered by the obliging media, these "camping trips" invariably ended with an open invitation to visit the Blackfeet in the new national park. In nearly every advertisement and press release, Great Northern publicists referred to the Blackfeet as the "Glacier Park Indians" and often encouraged visitors to come and acquaint themselves with these "specimens of a Great Race soon to disappear." Such appeals made good use of the "vanishing Indian" sentiment that tugged at the heartstrings of many Americans, and early visitors responded by making the Blackfeet an important part of their "wilderness experience."[2]

The Glacier Park Indians. Photographs by Roland Reed used in the magazine *Travel* (May 1914). The visual association of the Blackfeet with Glacier's dramatic vistas proved an important aspect of Great Northern Railroad's promotion of the national park. The upper image, entitled *The Watering Place,* was staged by the photographer to depict "a truly primitive and beautiful American scene, devoid of any element that savors of present-day civilization, and full of the poetry of the great outdoors and of a vanishing race." Below, *An Indian Chief and His Daughter* presents two more individuals staring off into the scenic distance. Here, the Blackfeet have "all the trappings and ornaments that one usually associates with the story-book redskin. Among such glorious natural surroundings as this, he and his women make a picture that is irresistibly charming." (Courtesy of the William Andrews Clark Memorial Library, University of California, Los Angeles.)

Blackfeet boy with golfer at Glacier Park Lodge. As "primitive" accouterments to the "civilized" splendors of Great Northern Railroad's luxurious hotels, the Blackfeet proved an essential aspect of the tourist's experience in Glacier National Park. Against the backdrop of spectacular mountain scenery and literally a stone's throw from a luxurious train depot, tourists could view Indian dancers, visit a tepee camp, golf, swim, and play tennis at the Great Northern Railroad's massive Glacier Park Lodge. As this image suggests, the so-called Glacier Park Indians not only "invited" Americans to come visit their homeland but also eagerly desired to entertain and serve them in any way possible. Yet such fictions only held within the context of recreational tourism, and park officials eagerly sought to exclude any sign of the Blackfeet from the Glacier backcountry. (Courtesy of the Glacier Natural History Association.)

Early tourists could hardly avoid associating the "Glacier Indians" with the park. From Indian art in the train cars to Blackfeet greeters at the railroad stations, from tepees and dances on the manicured lawns in front of Great Northern's magnificent hotels to the veritable museum of Plains Indian artifacts on display in hotel lobbies, they must have felt surrounded by Indians.[3] As early park promoter Edward Frank Allen wrote, Glacier seemed to preserve a place and a time when "the Indians knew not the restrictions of the reservation." Consequently, "the majesty of the mountains [remains] as undefiled and as poignant as [ever], and the region is still aloof from the desecrating hand of man."[4]

Blackfeet dancers at the hotels certainly imparted a sense of Indian presence to the mountains beyond, but their physical absence from the most popular sights in the park actually enhanced the tourist experience of Glacier. Allen reminded his readers that the Blackfeet may have "hunted in the mountains and fished in the lakes [but these] are now yours as an American citizen. That is a better thought as you glide over the surface of [a] lake than it possibly can be as drawn on a printed page. Glacier Park is yours and your children's!" Robert Sterling Yard echoed these sentiments in a popular souvenir publication on the national parks that he wrote with the full support of Secretary of the Interior Franklin K. Lane and Director of the National Park Service Stephen T. Mather. As Yard put it, "Glacier [was] once the favorite hunting ground of the Blackfeet [but] . . . now . . . [it is] strictly preserved."[5]

The Call of the Mountains

Such a celebration of non-Indian wilderness did not stem from a latent fear of past "atrocities." Instead, it reflected the deeply held values that shaped and defined how Americans experienced wilderness in Glacier National Park. Located in the northernmost stretch of the American Rockies, Glacier was both figuratively and literally the crowning natural feature of the United States. Along with Yosemite and Yellowstone, the park instantly became one of the nation's most "sacred places," where tourists combined an experience of sublime nature with a deep sense of patriotism. The Blackfeet may have used the Glacier area in the past, but tourists and park managers believed that only the citizens of an emerging world power could experience the mountains with appropriate awe and reverence.[6]

As "undiscovered" or rarely visited portions of Blackfeet territory, the northern Rockies once had intrinsic importance for a few thousand Indians; as the "Crown of the Continent," Glacier represented the power and grandeur of the whole United States. Furthermore, the area's designation as a "public park or pleasuring ground for the benefit and enjoyment of the people" reaffirmed the most cherished principles of democracy and equality.[7] Of course, the Blackfeet could also enjoy the park, once they had become sufficiently civilized and embraced all the prerogatives and benefits of citizenship. In the meantime, Glacier National Park belonged to "the people" while the Blackfeet belonged on their reservation, where they could best pursue the "white man's road."[8]

Although a visit to the new national park certainly appealed to the patriotic sensibilities of many tourists, particularly during World War I, the idea of spending time in the Montana wilderness tended to draw visitors at a much more personal level. "The Call of the Mountains," as Mary Roberts Rinehart described it in an often reproduced promotional piece for the Great Northern Railroad, offered the promise of throwing "off the impedimenta of civilization [and] the lies that pass for truth." In the mountains, you could "throw out your chest and breathe; look across green valleys to wild peaks where mountain sheep stand impassive on the edge of space" and experience your "real" self.[9] Such language had strong appeal to the well-off, native-born Americans who made up the majority of Glacier's first tourists. For them, the national park offered a refuge from the profound social and political changes that characterized early-twentieth-century America. Unable to influence the effects of mass urbanization, immigration, and industrialization and seemingly divorced from the manners and traditions that once defined American life, these early tourists came to Glacier for an "intense experience" that would give meaning to their increasingly complex and impersonal lives.[10]

Ultimately, the personal needs of "overcivilized" tourists found partial expression in a popular definition of the American past, one that promised to revitalize modern life and reclaim the older republican ideals of self-sufficiency and civic virtue.[11] In a very powerful sense, Glacier presented a fantasy realm where individual Americans could play out little frontier dramas and, like their European forebears, reinvigorate their lives through contact with the essential elements of the American wilderness. Native people inhabited the fringes of this frontier, standing at train stations or dancing in front of the hotels, but they remained wholly absent from the tourist's experience of the "real" wilderness. Outside the hotels, Glacier National Park preserved a vestige of the "virgin" continent, where vacationing adventurers could experience the North American wilderness at the dawn of "discovery." In short, the perceived absence of Indians from the Glacier backcountry made the park a virtual tabula rasa, on which tourists could freely project some of their most basic needs and desires.

Though men and women shared similar views about Glacier's "virgin" landscapes, they derived slightly different benefits from their wilderness experiences. For Theodore Roosevelt and his male companions, the supposed frontier conditions preserved in places like Glacier stimulated "that vigorous manliness for the lack of which . . . the possession of no other qualities can possibly atone." As Mary Roberts Rinehart pointed out, however, the Glacier wilderness also provided an arena where "the nervous woman" could conquer her fears and swell "with pride and joy." In other words, wilderness adventure allowed her to exercise the prerogatives of the so-called New Woman. Like Rinehart, Ada Chalmers learned that "women can get along in the parks without men, and in this way [Glacier] opens up a new and immense field of enjoyment." Lulie Nettleton, who traveled with the Mountaineers hiking club of Portland, Oregon, met a party of five women on the trail who "especially impressed [her]. They told of their wonderful experience in Glacier Park and the ease and safety with which women can travel alone and manage their own expedition." While personally liberating, these adventures also seemed to have a larger social importance. As one German man noted upon seeing the same group of women: "They could not do it in the Fatherland." No, they could not, and therefore all the more reason for Nettleton and her companions to support the current war against the Kaiser.[12]

Naturalizing the Wilderness

While tourists viewed Glacier as a national symbol, grand pleasure resort, and personal proving ground, park officials were obligated to maintain this uninhabited wilderness preserve to meet two other significant objectives. The legislation that created Glacier not only made note of the region's scenic beauty and recreational potential but also emphasized the area's importance as a game preserve and arena of scientific inquiry. Because the park contained a giant laboratory of wildlife, glaciers, forests, and rare flora, the Glacier National Park Act stipulated that, in the interest of science and game protection, all must remain in an undisturbed "state of nature."[13] An abundance of wild animals in the park also furthered the recreational and wilderness appeal of Glacier because tourists expected to see wildlife near the roads and hotels and in the backcountry. Nevertheless, Glacier required a great deal of management and manipulation to keep it in its "original" wilderness condition, and park officials implemented programs of predator reduction and winter feeding to increase the numbers of deer, mountain goats, and elk.[14]

During the early years of the park's administration, ungulate populations swelled after rangers and licensed hunters poisoned or killed hundreds of coyotes and dozens of eagles, mountain lions, and wolves.[15] These efforts helped to increase the numbers of "preferred" species in Glacier National Park, and elk populations were further increased by the importation of dozens of animals from Yellowstone in 1912.[16] To improve the chance that tourists might see these animals on a regular basis, park administrators also made plans to sow hayseed in some of the more frequently visited areas of the park and considered placing salt licks near popular scenic vistas. More significantly, Glacier officials devoted a great deal of energy and time to trying to eliminate hunting within the park. Killing game animals not only violated the basic principles that led to the establishment of the national park but also seemed to directly undercut the varied efforts to manage and augment Glacier's special wilderness qualities. By 1914, park officials could boast of great success against poachers from southern Alberta and the communities to the west of the park. Similar claims could not be made regarding hunters from the east, however, where the park bordered the Blackfeet reservation.[17]

No doubt chafing at any mention of the Great Northern's "Glacier Indians," park administrators made repeated attempts to exclude native hunters from the Glacier backcountry from the early 1910s onward. Problems with the Blackfeet involved more than just concerns about game protection. As Stephen T. Mather informed the secretary of the Interior in November 1915, treaty rights could potentially undermine the federal government's ability to exercise "exclusive jurisdiction over the park."[18] Because the 1895 agreement specifically guaranteed the right to hunt, and park officials placed so much emphasis on their mandate to manage game animals, hunting became the focus of the government's opposition to native use of park lands. Nevertheless, any Blackfeet claims on Glacier directly challenged the authority of the Department of Interior and seemed to undermine the recreational, preservationist, scientific, and symbolic importance of the national park.

The issue of Blackfeet rights in the park would quickly spread beyond the scope of a local dispute between rangers and native hunters. In time, it became a protracted battle that involved various federal and state agencies, touched on key constitutional issues, and captured the attention and energy of leading preservationists. Of course, the idealization of uninhabited wilderness affected others besides the Blackfeet. Exclusion of Indians from Glacier's backcountry accompanied similar efforts to prevent the killing of game by non-Indians.[19] Likewise, park officials sought to extinguish all private inholdings in Glacier—a problem that plagued the management of most parks at the time. Although privately owned land within the park's boundaries could be purchased by the government, as most was by the 1930s, and white hunters quickly learned to avoid Glacier, the Blackfeet continued to press their claims to the eastern half of the park. Consequently, the issue of Blackfeet rights would outlast and far surpass any concerns about white hunters and private landholders.[20] The legal complexities of the conflict that arose between the park service and the tribe and the radically different meanings that Indians and non-Indians attached to the park landscape created a unique and potent arena of contention that continues to defy easy resolution.

Land Wars

Beginning with the first two annual reports to the secretary of the Interior, Glacier's early superintendents emphasized the need for more personnel to enforce game protection laws, particularly along the park's boundary with the Blackfeet reservation.[21] Rangers frequently encountered native hunters on the eastern side of the park, and questions soon arose about Blackfeet rights in Glacier. In the summer of 1912, the acting superintendent wrote a circular report for park staff and the agent at the Blackfeet reservation, explaining that Indians had no rights to hunt in the park.[22] Whatever the agent may have told them in regard to these matters, Blackfeet hunters continued to pursue game in the mountains without hesitation.

In his annual report for 1913, Glacier Superintendent James Galen complained of Blackfeet hunters over the previous two years and, as he put it, "most emphatically recommend[ed] an extension of the park on the eastern side" to push back Indian hunters and prevent them from "ruthlessly slaughter[ing]" elk and other game animals.[23] Nothing immediately came of this idea, but concerns about native hunters did reach the commissioner of Indian Affairs in the summer of 1915. As if to make up for two years of inaction, he immediately directed the Blackfeet agent to inform the Indians that they had no rights to hunt in the national park.[24] The agent quickly posted a general notice throughout the reservation, but it could not have come at a worse time.[25] Already "very much agitated over" rumors about a possible park expansion, the notice only served to inspire an outpouring of protests regarding any infringement of Blackfeet rights in the mountains.[26]

At least a third of the tribe depended to some extent on the resources of the Glacier region in the park's early years, and still more felt upset about any challenge to their sovereignty.[27] Following the posting of the notice against hunting, a Blackfeet man named D. D. LaBreche wrote Montana Senator Harry Lane about tribal rights in Glacier. Enclosing a copy of the 1895 agreement, he informed Lane that the Blackfeet never relinquished these rights. With a keen sense of the language contained in both the earlier agreement and the legislation creating the national park, he asked: "Does the same remain 'Public Lands' of the United States? If not, who does it belong to?" He could only conclude that any effort to deny the Blackfeet access to the park was "an act of injustice"; at the very least, the tribe deserved compensation "for the privilege which they retained, and never sold to the government."[28]

LaBreche's letter to Lane was followed by another from Peter Oscar Little Chief, the secretary of the Blackfeet Tribal Council, who also inquired about his rights to hunt and collect wood in the park.[29] Together, LaBreche and Little Chief set in motion a flurry of activity and concern in Washington. Senator Lane's office forwarded both letters to the commissioner of Indian Affairs, who in turn informed the secretary of the Interior. Especially troubled by the arguments contained in LaBreche's letter and apparently unsure of the legal authority behind earlier orders regarding Blackfeet hunting in the park, then Assistant Secretary of the Interior Stephen T. Mather requested an opinion from Interior Department attorneys.[30] By early January, government solicitors had written an eight-page report in which they concluded that tribal rights ceased with the creation of the national park. The secretary directed that a copy of this report be sent to the Blackfeet agent and ordered him "to give due notice thereof to the Indians under his charge."[31]

Although the Glacier Park act remained silent on the question of Blackfeet rights, Solicitor Preston C. West concluded that park administrators had good legal grounds to exclude Indians from Glacier. Based on a close reading of the Supreme Court's decision in *Ward v. Race Horse*, the solicitor's opinion rested on three key points: because a national park is not subject to disposal or sale by the federal government, it cannot be considered public land; the creation of the national park directly undercut the 1895 agreement because Blackfeet rights in the Glacier area held only so long as it remained "public land of the United States"; Congress had plenary authority over Indian affairs, which gave it the right to abrogate the earlier agreement by creating a national park and thereby changing the "public" status of the lands in question. While these legal conclusions must have brought some comfort to park administrators, LaBreche, Little Chief, and others carried on their protests to government officials. Still more Blackfeet continued to exercise their rights, as they interpreted them, within Glacier National Park.[32]

Along with concerns about Indian hunters on the eastern side of the park, Glacier officials became deeply troubled about their inability to control hunting on the reservation as well. Having taken great pains to increase the numbers of game animals, the park service felt a proprietary interest in the elk and deer that moved onto the reservation during the fall and winter months. Beginning in the summer of 1917, Glacier officials sought to implement a series of legal restrictions that would prevent the Blackfeet from hunting any migratory animals that "originated" in the park. They appealed to the Bureau of Indian Affairs for help, but the commissioner declined to support these efforts; as he repeatedly informed Glacier Superintendent Thomas Ferris and park service officials in Washington, all Indians had a sovereign right to hunt on their reservations free of outside regulation.[33]

Unable to restrict hunting on Blackfeet lands, park officials then tried to resurrect the earlier proposal to move the park boundary further eastward. Through a complex and difficult process, the park service hoped to buy up 22,000 acres of individually allotted land and 37,000 acres of tribal timber reserves and reclamation withdrawals. These purchases would then require the removal of hundreds of individuals from their homes.[34] Though a legal and logistical near impossibility, park officials believed the importance of incorporating reservation lands into the national park could not be underestimated. If successful, the plan would not only push the Blackfeet further from the heart of the park, thus making it more difficult for Indians to enter the mountains, but also bring the entire winter range of migratory game animals under the authority of the newly created National Park Service.[35] As Acting Supervisor George E. Goodwin put it in his annual report for 1917, extending the park six miles into the Blackfeet reservation would allow game animals to roam unmolested within their "natural boundaries."[36]

Both the commissioner of Indian Affairs and the agent for the Blackfeet reservation dragged their feet on these later proposals, but they could not long withstand a growing chorus of demands for some resolution of the "Blackfeet problem."[37] Concerned leaders of national preservationist movements got wind of Glacier's troubles with the tribe and demanded that Secretary of the Interior Franklin K. Lane put a stop to the "wanton slaughter" of elk. William T. Hornaday, the director of the New York Zoological Park and trustee for the Permanent Wild Life Protection Fund, complained of "Piegan Indians . . . openly slaughtering the game in the northern portion of Glacier Park" and urged the secretary to force the Bureau of Indian Affairs to "take steps to remedy this abuse." Horace Albright, then assistant director of the National Park Service, joined Hornaday in berating the Indian Service and wrote that "anybody who knows anything about this feels that if the Indian office were drastic enough in its control the killings could be stopped." Even the venerable George Bird Grinnell lent his considerable weight to the issue and suggested that "this matter may easily be remedied provided the Agent is willing to give orders to his Indians that the laws of the State of Montana are in operation on the Indian reservation." Although such laws had no authority on tribal lands, Grinnell felt that enough pressure from Secretary Lane could persuade the Indian agent to enforce state game hunting seasons on the reservation.[38]

Hornaday followed up on Grinnell's advice and, in a subsequent letter to Lane, reminded the secretary about the seriousness of the entire affair. "In view of the character of Glacier National Park and all that it stands for with the American people," he wrote ominously, "I think it would be deplorable for any of its winter driven game to be killed for food contrary to the laws of the State of Montana." Although Lane complained of the difficulty in policing the western boundary of the reservation, he conceded "the importance of affording every protection to the game of our National Parks" and promised that "a special effort will be made to see that there is no further abuse by the Indians of the Blackfeet Reservation."[39] A newly appointed agent for the Blackfeet was briefed on the matter and directed to cooperate in "the fullest measure . . . with the authorities of the Glacier National Park." He was then ordered to post a notice throughout the reservation with the threat that "any Indians who persist in killing [Glacier National Park] animals will be prosecuted" by the Court of Indian Offenses.[40]

As before, threats and warnings against hunting "cause[d] considerable discussion, and resentment among the Indians."[41] Still, the Blackfeet largely ignored

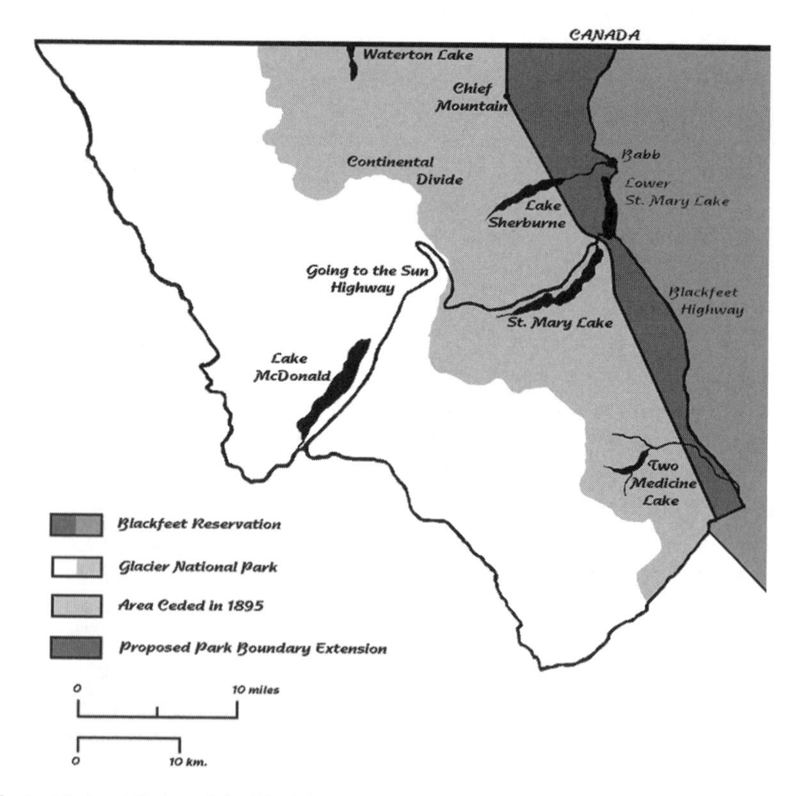

Glacier National Park and the Blackfeet reservation.

these restrictions, and within a few years a new commissioner of Indian Affairs once again found himself directing the Blackfeet agent to curb Indian "abuses." By the early 1920s, however, no one on the reservation could afford to remain cavalier about these periodic directives from Washington. Prior to this time, the issue of Blackfeet rights in the park tended to concern only the families who lived closest to the area. Though many who did not go into the mountains shared the sentiments of older leaders like Tail Feathers Coming over the Hill, who despised both park officials and their regulations, those who continued to hunt or gather in the park generally avoided ranger patrols and managed to use the area much as they had in the past.[42] But cooperative efforts between park rangers, state game wardens, and reservation officials to exclude Indians from Glacier and even restrict on-reservation hunting of animals that moved back and forth across park boundaries were finally proving successful. Consequently, as native use of the park dropped off, concerns about tribal rights became more widespread throughout the entire reservation community.[43]

Chafing under these new restrictions, scores of Blackfeet expressed their frustrations to Montana Senator Thomas Walsh. Peter Oscar Little Chief, whose inquiry about Blackfeet rights in Glacier had troubled park officials some ten years previously, wrote a petition in the fall of 1924 and gathered pages of signatures asking Walsh to introduce legislation specifically guaranteeing Indian rights to hunt in the eastern half of the national park. The men and women who signed Little Chief's petition realized that trouble with the park service had spread far beyond an objectionable notice from the agent or the occasional encounter with a ranger. Whether they hunted in the park or not, all realized that efforts to exclude them from Glacier and restrict hunting on their reservation impinged directly upon the political sovereignty of the entire tribe.

Little Chief made plain that, according to the Blackfeet, the tribe had "sold the United States Government nothing but rocks only. We still control timber, grass, water, and all big game or small game or all the animals living in this mountains. The [agreement of 1895] reads that as long as the mountains stand we got right to hunt and fishing. And provided further that the said Indians hereby reserve and retain the right to hunt upon the said lands . . . so long as the same shall remain public lands of the U.S."[44] Sent on to the Bureau of Indian Affairs, the petition was apparently ignored and ultimately disappeared from bureau records altogether. Having never received a reply, Little Chief again sent letters to Walsh in 1926 and 1928 to inquire about the fate of the petition. In 1928 the commissioner of Indian Affairs finally responded but sidestepped the question of specific rights to hunt in the park. As he told Walsh, who forwarded the letter to the Blackfeet leader, "outside the reservation, the State game laws apply to the Indians, who are now citizens, as well as to the whites and other citizens of the State." Consequently, the Blackfeet had the same rights to hunt in Glacier National Park as non-Indians—which is to say, no right at all.[45]

The dismissive nature of the commissioner's response certainly upset Little Chief and probably contributed to the growing tension between park rangers and Blackfeet hunters along the front range. While economic changes on the reservation had made the mountains less important to a growing number of Blackfeet, and many began to abandon the park altogether, others viewed efforts to exclude them from the eastern half of Glacier as a personal affront. Chastised by reservation agents to obey state game laws while on the reservation, and subject to arrest and heavy fines by a growing force of park rangers, some chose to lash out. By the early 1930s, a near state of war existed on the eastern side of the park, with Blackfeet and rangers prepared to shoot and be shot upon at any given time. A siege mentality soon developed in Glacier, and park officials sent out defensive scouting parties to report on the movements of "Injun hunters," "prowlers," and "gut eaters" from the Blackfeet reservation.[46]

Although a diminishing number of angry individuals continued to defy park regulations, their actions received widespread support throughout the reservation. When four men were arrested for hunting in the park in October 1932, they chose to fight their case based on the rights reserved to the Blackfeet in the 1895 agreement. Though convicted and fined the following month, they hired attorneys and appealed to the U.S. District Court in Helena, Montana. In support of their appeal to the higher court, a number of dances and other events were organized on the reservation to raise money to fight the "case all the way to the Supreme Court if necessary."[47]

Widespread interest in the case was no doubt fueled by the efforts of the tribal council to extend the reservation's western boundaries four miles into the park. In response to their many conflicts with Glacier officials, a number of Blackfeet had poured over the official minutes of the 1895 agreement. Finding an apparent inconsistency in the original surveyor's notes, they promptly informed the tribal council, which then petitioned the secretary of the Interior for the establishment of a new boundary between the reservation and the national park. Of course, this proposal had tremendous appeal across the reservation. Few could resist the terrific sense of comic justice that mocked the park service's ongoing plans to enlarge the eastern side of the park.[48]

Nothing ever came of this attempt to reduce park lands, but park officials began to wonder if they would ever resolve their problems with the tribe. In the spring of 1932, Glacier Superintendent Eivind T. Scoyen confided to Horace Albright, now director of the National Park Service, that he fully expected "to go to court on this matter . . . [but despaired] that legally we [may] have no right to prevent the Indians from [using the park]."[49] National Park Service officials in Washington shared Scoyen's concerns and promptly forwarded his letter to the secretary of the Interior to request a new ruling on Blackfeet rights in the park from the Office of the Solicitor. Before they could receive this much anticipated legal report, the district court agreed to hear the appeal from the four Blackfeet hunters. Scoyen realized the gravity of the situation and, along with Mather and the secretary of the Interior, appealed directly to the attorney general for assistance. For park service officials, no matter could be of greater importance. "This is an extremely important case as far as we are concerned," Scoyen told the solicitor's office, "and no effort should be spared to get a favorable decision for the Government."[50]

Even more ominous than the appeal before the district court was a pending decision in the U.S. Court of Claims that could force the government to recognize Blackfeet rights in Glacier and pay restitution for their temporary obstruction. Part of a case originally begun in 1909 by Little Dog, the most obstinate negotiator during the 1895 land cession negotiations, the Blackfeet had joined with other tribes to seek restitution from the United States for a series of treaty abuses that extended all the way back to 1855. As the case moved through the U.S. Court of Claims, however, tribal lawyers began to focus attention on the 1895 land cession agreement and current Blackfeet rights in the national park. Park service officials were certainly aware of the strong evidentiary testimony detailing Blackfeet use of the Glacier area before 1910, and Scoyen must have felt as if he was about to preside over the loss of one of the nation's most prized crown jewels.[51]

Stalemate

In the midst of these worries, both Scoyen and Director Albright took comfort in the findings of the solicitor's report they had previously requested from the Department of the Interior. Basing his argument on a narrow definition of public lands, Solicitor E. C. Finney concluded that Glacier National Park represented "lands set aside for a public purpose. When the lands were disposed of or reserved for a public purpose, they ceased to be public lands of the United States as the term is understood in legislation." Perhaps aware that his line of reasoning might prove unsatisfactory, Finney then noted that the 1895 agreement made Blackfeet rights to the mountains contingent on the game laws of Montana. When the state ceded its jurisdiction over the park, Blackfeet rights became subject to federal authority. If he had known that tribal leaders were probably unaware of the clause that subjected them to Montana's game laws, Finney might have cited the case of *Ex parte Crow Dog* (1883) and ruled that Blackfeet understandings of the agreement must take precedence over the written document. But like Solicitor West before him, Finney also based his decision on the precedent established in *Ward v. Race Horse.* Consequently, he determined that Blackfeet "privileges . . . were terminated when Congress exercised its power to reserve the lands and dedicated them to a particular public use."[52]

Scoyen made good use of the new report, forwarding a copy to the assistant U.S. attorney handling the district court case. It eventually proved an effective cornerstone of the government's argument, and the court upheld the previous conviction of the Blackfeet hunters.[53] The legal position of the National Park Service vis-à-vis the Blackfeet was even further strengthened by the 1935 Court of Claims decision in the case of *Blackfeet Indians (et al.) v. United States.* The court did rule in favor of the Blackfeet with regard to past treaty violations on the part of the United States and awarded some financial compensation to the tribe. In regard to the 1895 agreement, however, the court made special note that Blackfeet rights within Glacier National Park deserved neither recognition nor compensation. As the final ruling stated, the Blackfeet "did not exercise to any appreciable extent the rights reserved" in the 1895 treaty, and thus the "rights were terminated" by the Glacier Park Act of 1910.[54] Given all the concern that park officials had expressed over Indian hunters, the court's decision must have seemed a bitter irony to tribal leaders.

Just as the park service reached its strongest position relative to Blackfeet claims on Glacier, the old dream of extending its authority over reservation lands was beginning to slip away. At least since 1919, when William T. Hornaday and George Bird Grinnell complained of the "wanton slaughter" of elk, Director of the National Park Service Stephen T. Mather had taken a personal interest in plans to extend the park several miles into the Blackfeet reservation.[55] As noted previously, the Bureau of Indian Affairs declined to support these plans. When concerns over Indian hunting increased in the late 1920s, however, park officials once again renewed their efforts to extend Glacier to its "natural eastern boundary."[56] In the midst of Peter Oscar Little Chief's petitions against Blackfeet exclusion from the park, then Assistant Director Horace Albright elicited the support of Great Northern President Ralph Budd for the eastward extension of the park.[57] These efforts apparently fell on deaf ears in Washington, but Albright returned to the issue once he assumed the directorship of the National Park Service from the ailing Mather in 1929.

After sustained pressure from the new director, Commissioner of Indian Affairs Charles Burke agreed to give weighty consideration to Albright's concerns. In January 1930, Burke admitted "that from the viewpoint of the Park Service, the proposed extension of Glacier Park boundaries . . . would be . . . in the very best interests of the public generally," but he could give only qualified support. Aware of previous Blackfeet protests regarding earlier proposals to enlarge the national park, he insisted that the matter must be put directly to the Indians themselves.[58] At a tribal council held the following March, Blackfeet leaders were emphatically and unanimously opposed to the proposal.[59] At a less formal meeting with the assistant secretary of the Interior a few months later, the issue came up again. A venerated Blackfeet elder named Mountain Chief spoke for many when he condemned the park service's efforts to have "a six mile strip of land on the reservation to be added to the national park. Just forget that idea. I said no. We don't want to part with it. . . . They said they are to going to put buffalo on the land [but] . . . I think it is just a bait."[60]

Perhaps still believing that a small bison herd might do the trick, Albright hoped the Blackfeet might be swayed at a later date. He seemed to sense a new opportunity after a surprisingly pleasant meeting with Joe Brown, president of the Blackfeet Tribal Council, and two other council members at his Washington office in February 1931. Brown and the others had come to ask the director to do all in his "power to secure employment for Blackfeet Indians in Glacier National Park." While he made no promises, Albright found the "Indians . . . very friendly" and believed they felt "the National Park Service wants to be as friendly as possible to the Blackfeet Indians." Based on these impressions, Albright immediately ordered "a review of all the correspondence in regard to our desire to [extend] the east boundary."[61] By year's end, he had drawn up specific legislation for the eastward expansion of Glacier National Park and appealed to a new commissioner of Indian Affairs for support.[62]

Like Albright, the commissioner was aware of the two legal cases involving Glacier National Park and the Blackfeet. Unwilling to exacerbate matters any further, he followed the lead of his predecessor and once again insisted that his support of the legislation would depend on Blackfeet acceptance.[63] Following the commissioner's instructions to assess Blackfeet attitudes toward a proposed park expansion, Blackfeet Agent Forrest Stone met with the tribal council "and some fifty or seventy-five other Indians that were in attendance." Their angry response was unanimous and unequivocal. Consequently, Stone informed the commissioner "that the introduction of [Albright's] bill would be the most unpopular proposition that could ever be presented to the entire tribe. [Probably no more than] twenty Indians of the entire tribe of 3700 would give their consent to the passage of such a measure." Stone concluded his report with a strong recommendation that the "proposed bill be withdrawn and that the Park Service be given to fully understand the attitude of our Indian people, in the hope that no further steps will be taken in this direction."[64]

Blackfeet view of park expansion, 1930. By unanimous decision of the Blackfeet Tribal Council, this cartoon by Richard Sanderville was included in a report to the Commissioner of Indian Affairs. It depicts a sharp-clawed Horace Albright, Director of the National Park Service, grasping toward lands on the Blackfeet reservation. Though humorous, it clearly represents the hostile nature of Blackfeet–park service relations in the early 1930s. (NA RG 75, BIA CCF, 1907–39, Blackfeet 307.2/59976; Courtesy of the National Archives, Washington, D.C.)

In the words of historian Hal Rothman, Albright had "piranha-like instincts" when it came to "the politics of land acquisition."[65] He also seems to have possessed a remarkable degree of single-minded determination. Unable to gain the support of the commissioner of Indian Affairs, Albright's bill died a quiet death in congressional committee. Still, he never let go of his dream to extend the boundaries of Glacier National Park. Although he retired from the park service in 1933, he apparently shared his concerns with his successors—who found a new opportunity to broach the issue when the court of claims ruled against the Blackfeet in 1935. Perhaps wary of the Bureau of Indian Affairs, park service officials decided to bypass the commissioner and approached the Blackfeet directly. When Superintendent Scoyen met personally with Blackfeet Agent J. H. Brott and the Tribal Business Council, he learned firsthand just how much bad blood there was among the Blackfeet in their opinions of the National Park Service. Much to his regret, he had to inform his superiors in Washington that it was "useless for the NPS to initiate any action or to support any project which has for its objectives the gaining of control over lands in the so-called Blackfeet strip." Officials in Washington heeded Scoyen's advice and no longer gave serious consideration to the extension of the park's eastern boundary.[66]

By 1935, relations between the Blackfeet and the National Park Service had reached an impasse that remains unresolved to this day. On one side, the park service, tourists, and preservationists had largely made Glacier into the uninhabited wilderness that continues to inform potent ideas about nature and national identity. Nevertheless, Blackfeet use of park lands undermined this idealized notion of wilderness, and the tribe's united resistance to Glacier's expansion set limits on its physical expression. As relations between the Blackfeet and the park service entered stalemate, tribal leaders embarked on an unofficial policy of noncooperation. Though individuals continued to entertain tourists in the park, gone were the days when whole delegations would lend their photogenic presence to an official park celebration. More significantly, the tribal council generally ignored overtures to improve relations or work on joint conservation efforts. For the Blackfeet, such collaboration would have indicated a recognition of the park service's authority over the lands ceded in 1895.

The Blackfeet also took advantage of the so-called Indian New Deal, a series of federal policies designed to halt the loss of any more reservation lands. Becoming one of the first native groups to form a tribal government along the lines of the Indian Reorganization Act, Blackfeet leaders built a strong alliance with Commissioner of Indian Affairs John Collier just as park officials were about to make their final pitch to buy tribal lands. When Superintendent Scoyen met with the Blackfeet in April 1935, he addressed one of the first meetings of the newly reorganized tribal council. Their adamant rejection of his plans reflected more than bad blood and drew on more than the active interest of the new commissioner. Blackfeet resistance was now rooted in a strongly centralized body of political authority. As Scoyen was shocked to learn, negative opinions of the park service had become a central aspect of tribal policy and a fundamental expression of Blackfeet national identity.[67]

The Blackfeet may have hardened in their dealings with the park service, but Glacier's environment changed in ways that eased tensions between the two parties and allowed matters to settle into a protracted truce. Park officials no longer concerned themselves with the loss of game animals in the late 1930s but worried instead about the overgrazing of shrubs and grasses. Years of winter feeding programs and predator reduction had pushed the populations of deer and elk beyond the carrying capacity of the park's eastern slope. At one point, feeding programs were shut down completely in an effort to starve "surplus" game animals.[68] By the mid-1940s, Glacier officials even went so far as to condone Blackfeet hunting on the western edge of the reservation and probably turned a blind eye to Indian hunters within the park because "year-round hunting for Indians [would give] enough annual take" to keep game populations in check.[69] Within a decade, however, overgrazing had become such a serious problem in the park that Superintendent John Emmert complained of Blackfeet hunters not killing enough animals.[70]

Although tensions between the tribe and the park service subsided, the issue of Blackfeet rights in the eastern half of Glacier would not disappear. In the late 1940s and early 1950s, a number of tribal elders who had been present at the 1895 negotiations began to inquire about when the government would return the eastern half of Glacier National Park to the Blackfeet. Many claimed that the land cession only amounted to a lease of fifty years, and time had run out. Though some points in the transcribed negotiations do mention a fifty-year moratorium on further government requests to purchase more reservation lands, Blackfeet understandings of a terminal lease were apparently lost in translation.[71] Inspired or perhaps cajoled by these elders, the tribal council brought two new cases against the government in 1954 and again in 1966. The fading memories of tribal elders could not stand against written documents, however questionable the accuracy of their translation and transcription, and the Blackfeet lost both cases.[72]

By the 1960s, few Blackfeet actually hunted near the park and even fewer went to the mountains to gather traditional plant foods and medicines. But the continuing importance of the Backbone of the World did not depend on how many people went to the mountains. In 1895, the Glacier region had certainly provided the tribe with a greater portion of its physical sustenance. Yet throughout seven decades of rapid economic and social change, the issue of Blackfeet rights in the area had always reflected basic concerns about cultural persistence and tribal sovereignty. This remained the case during the Red Power movement of the early 1970s, when a growing concern about traditional skills and practices throughout Indian country had a strong influence on Blackfeet leaders. Through direct action and protest, the tribal council and numerous individuals besieged Glacier officials with impassioned demands for immediate recognition of Blackfeet rights in the park.[73]

The new climate of Indian activism reinvigorated long-standing concerns, but it did not alter the basic tenets of past efforts to restore Blackfeet claims to the eastern half of Glacier. The near state of war that once characterized relations between the Blackfeet and park officials resurfaced in the early 1980s, and armed conflict was only narrowly avoided on several occasions.[74] In a more peaceful but no less hostile manner, the Blackfeet officially restrengthened their old policy of noncooperation and refused to allow the construction of a livestock fence along the eastern boundary of the park. While they acknowledged that such a fence would keep domestic animals out of fragile grasslands, council members worried more that it might indicate tacit approval of park authority.[75] Eventually, continued disagreements with the Blackfeet would force the National Park Service to revisit issues it thought had been buried in the 1930s.[76]

At the height of efforts to extend the park's eastern boundary, Horace Albright described the western portion of the Blackfeet reservation as a place that "is by topography, juxtaposition and character logically a part of Glacier National Park."[77] In many respects, he made a prescient argument for ecosystem preservation and management. Of course, the unstated logic was that Indians must be excluded for the ecosystem to operate "naturally." Although the proposal to enlarge the park at the further expense of the Blackfeet illustrates an audacity that has long characterized the park service's relationship with its Indian neighbors, it also marks a certain maturation of the American wilderness ideal. Albright's easy logic and the tenuous legal arguments that negated Blackfeet rights in Glacier demonstrate how completely linked preservation and Indian removal had become.

As the park service developed under Albright's forceful administration, the management of all parks became more regularized. Not surprisingly, Glacier provided an important model for policies of Indian removal at other national parks. Perhaps nowhere illustrates this better than Yosemite, where park officials began to view several decades of native habitation as a problem in need of a solution. Because of a number of unique historical conditions, Yosemite presented a remarkable exception to the general belief that parks and native communities could not coexist. The original park was established eight years before Yellowstone, yet native peoples became an integral part of Yosemite's early development. Like the Blackfeet at Glacier, Yosemite Indians were also important symbols of the national park in the 1910s—but this reflected decades of regular contact with tourists and long established residence within the valley. Nevertheless, Yosemite was an exception that would eventually prove a rule. Long after the mountain-dwelling Shoshone had been forced out of Yellowstone and just as the Blackfeet "problem" was becoming something of a moot point, park service officials began to implement a program to create a vision of pristine wilderness in the spectacular heart of the Sierra Nevada.

7

THE HEART OF THE SIERRAS, 1864–1916

> The power of scenery to affect men is, in a large way, proportionate to the degree of their civilization and the degree in which their taste has been cultivated. Among a thousand savages there will be a much smaller number who will show the least sign of being so affected than among a thousand persons taken from a civilized community.
>
> Frederick Law Olmsted, 1865[1]

AS THE EXAMPLES OF YELLOWSTONE AND GLACIER clearly demonstrate, Americans are able to cherish their national parks today largely because native peoples either abandoned them involuntarily or were forcefully restricted to reservations. For well into the twentieth century, however, Yosemite Valley remained home to a permanent, relatively autonomous Indian village. Whether cutting a trail on a government crew, working for a concessionaire, selling crafts to eager tourists, or providing information to a young anthropologist, the residents of the local native community made themselves an integral part of the national park—long after the dusty old days of land grabs and Indian wars. While native residence in the valley stands in marked contrast to other early parks, it also presents an important comparison with the experiences of Indian peoples at Yellowstone and Glacier. Such differences shed important light on Yosemite's unique history and reveal the processes by which this park was eventually made to fit the standards of the national park ideal.

The World Rushes In

The Yosemite Indians' ability to remain in a national park resulted in large part from a long history of efforts to both resist and adapt to the American conquest of their homeland. The first sustained contact between the Yosemite and whites took place in the midst of the Gold Rush, as thousands of Forty-niners invaded the central Sierra Nevada. In their feverish quest for some trace of the Mother Lode, miners brought epidemic diseases to native communities and destroyed carefully tended ecosystems. Moreover, the growth of mining camps and settlements also spawned a series of violent conflicts between whites and displaced peoples. Not surprisingly, the "discovery" of Yosemite Valley in 1851 occurred during a military campaign to subdue the peoples of the central Sierra Nevada and relocate them to the San Joaquin Valley. Efforts to remove the Yosemite Indians from the region ultimately failed, however, and they reestablished themselves in the valley after two years of sporadic encounters with miners and state militia battalions.[2]

By necessity, the Yosemite developed an accommodating relationship with nearby mining camps in the mid-1850s. Despite occasional flare-ups, Chief Tenaya endeavored to fulfill an 1852 promise to government officials that his people would avoid conflict with neighboring white communities. His efforts proved largely successful, and a number of Indians even started to work for individual argonauts or panned gold for themselves.[3] Yosemite Valley lay outside the purview of most mining interests, however, and the Indian community there managed to preserve a degree of distance and autonomy from neighboring white society that few native groups in the gold country could ever hope for. Consequently, Yosemite became something of a cultural island and remained, as it had been for centuries, an important place for hunting, harvesting various food and medicinal plants, and holding religious celebrations.

Only a small number of individuals remained in the valley year-round at this time, but hundreds left their winter camps in the lower country to the west and returned to Yosemite each spring. In 1857, for instance, an early hotelier observed that an especially "large band of Indians" had come to the valley "on account of a bounteous acorn crop the preceding fall."[4] A few weeks later, a Belgian gold miner familiar with the Yosemite region probably observed the same group of about one hundred when he noted that a large encampment he had encountered three years earlier had moved further up the Merced River into the valley.[5] Yosemite Indians still lit purposeful fires in the valley in the early 1860s, and one traveler observed that they had started so many for the purpose of "clearing the ground, the more readily to obtain their winter supply of acorns and wild sweet potato root," that the glow of the fires could be seen from miles away.[6]

The California Gold Rush took a severe toll on the people of the central Sierra Nevada, but native inhabitants still greatly outnumbered European and American visitors to Yosemite Valley until the early 1860s. Between 1855, when the first pleasure-seeking tourists visited Yosemite, and 1863, only 406 visitors entered the valley. As Yosemite's fame grew and travel became less arduous, however, visitation increased exponentially. In 1864, the year that President Lincoln signed the Yosemite Park Act, the valley received 147 visitors, but this figure more than doubled the following year and soon rose above 1,100 with the completion of the transcontinental railroad in 1869.[7] Along with increasing numbers of visitors, tourist facilities quickly expanded as early concessionaires built new hotels, planted orchards and vegetable gardens, plowed and fenced hay fields, blazed trails, and constructed roads.[8] Between 1874, when Yosemite received 2,711 tourists, and 1875, the Big Oak Flat Road, the Coulterville Road, and the Wawona Road opened to wagon traffic for the first time, bringing loads of supplies and coaches full of tourists to the valley on a regular basis.[9]

Despite the dramatic increases in visitation, Indians in Yosemite Valley remained on fairly good terms with their new neighbors. For the most part, they found in the growing tourist industry a means by which they could both earn a livelihood within their rapidly changing world and remain in their ancestral home. A number of small communities in the Sierra foothills made similar adjustments to the changes wrought by growing white settlements, but these *rancherias* generally persisted only as very small clusters of a few families and related individuals. The native population of Yosemite actually grew as tourism increased, however, and a number of dislocated groups returned to the area to seek employment during the spring and summer tourist season.[10]

How one defines a Yosemite Indian has long proven difficult for anthropologists and park officials, but the people most closely associated with Yosemite Valley in the midnineteenth century were the Ahwahneechee. Part of a larger cultural and linguistic group called the Sierra Miwok, the Ahwahneechee had lived in the Yosemite area for at least six hundred years. Whether they had replaced earlier inhabitants about 1100 to 1400 C.E., as archaeological and linguistic evidence suggests, or descended from people who settled in the valley some three thousand years previously, the Ahwahneechee viewed their home as the place where Coyote had especially directed them to live after the creation of the world. As a number of late-nineteenth-century Ahwahneechee related their history to outsiders, they described Yosemite as a special place the Creator had filled with all they would need, including trout, sweet clover, potent medicinal plants, roots, acorns, pine nuts, fruits, and berries in abundance, as well as deer and other animals, "which gave meat for food and skins for clothing and beds."[11]

Although they had once trusted to the remoteness of the valley to protect them from invading Americans, the Ahwahneechee were never an isolated people. They frequently traded and intermarried with other Miwok and with Mono-Paiute from the eastern side of the Sierra Nevada, a fact that may explain the present name of their home. In the Southern Sierra Miwok dialect, the Ahwahneechee are "the people of Ahwahnee," and Ahwahnee means "the place of the gaping mouth." Miwok people who lived west of the valley sometimes referred to the Ahwahneechee as *johemite*, which can be translated as "some of them are killers." Because most Sierra Miwok greatly distrusted the Paiute, this word probably refers to the presence of these people among the Ahwahneechee. The armed men who first entered the valley in 1851 must have learned of the Ahwahneechee village locations from these western neighbors, and so the present name of the valley dates back to this invasion.[12]

Besides the Mono-Paiute and other Sierra Miwok, Yokut from the Central Valley and some ex-mission Indians mixed in with the Ahwahneechee before the

1850s to create a complex Yosemite Indian society.[13] Such cultural blending was common among all precontact groups and generally followed long-established patterns of trade and exchange. These processes became less self-directed and more pronounced in the midnineteenth century, however, when native peoples struggled to survive the impact of American settlements. Lafayette Bunnell, one of the first whites to see Yosemite Valley during the militia campaigns of 1851 and 1852, clearly recognized all of these processes at work when he referred to the "Yo-Semite Indians [as] a composite race, consisting of the disaffected of the various tribes from the Tuolumne to King's River."[14] The processes of cultural blending, or ethnogenesis, did not cease with the end of the Gold Rush, and Yosemite Indian culture continued to evolve in the decades following the establishment of Yosemite Park. Borrowing items and practices from surrounding American and Mexican communities and combining the traditions of various Indian groups, the Yosemite constantly adapted to new conditions and managed to remain a distinct and viable community.[15]

Although they retained a fair amount of their older cultural practices, the Yosemite became further integrated into the tourist economy as more and more visitors arrived in the valley. Increasingly, the Indians' presence in the valley depended on their ability to gain employment from hoteliers and concessionaires. Men found work chopping wood and putting up hay, labored about the hotels, served as guides, drove sight-seeing wagons, and often provided large private parties with fish and game.[16] The Yosemite succeeded especially well at supplying fish to tourist parties, who, as many sportsmen reported, almost never had any luck fishing. As one early visitor noted, "Trout are abundant in some of the streams, but they are very shy of the hook. The Indians catch them in traps, and frequently supply travelers at twenty-five cents per pound."[17] Yosemite women often worked in the private homes of concessionaires as domestics, and in the hotels they found work as maids or laundresses.[18] Women and children also picked the wild strawberries that grew in the valley meadows in late summer and sold them to the hotels, and even as late as 1913 private parties could still occasionally purchase chickens, fresh fish, and wild strawberries from Yosemite families.[19]

Native employment in Yosemite reflected patterns established throughout the Sierra Nevada in the years following the Gold Rush. The massive invasion of miners who poured over the mountains brutally displaced entire native communities, while the environmental destruction wrought by mining practices undermined seasonal hunting and gathering cycles. Severely weakened and suddenly homeless in their homelands, most of California's shrinking native population found the means for survival only in close accommodation with non-Indian society.[20] Many Miwok families and individuals moved to where they could eke out a living on the margins of white settlements. Though generally despised and frequently humiliated by whites, their presence was tolerated whenever native labor could not easily be replaced by Mexican or Chinese workers.

A similar situation developed in Yosemite, but there native people got along much better with their non-Indian neighbors. Although a Yosemite man named

Choko complained in the mid-1870s that "white man too much lie," at least the valley did not attract the same rough crowd that congregated in other parts of the Sierra Nevada.[21] The remoteness of Yosemite also made native labor more prized, and because they posed no visible threat to tourists or concessionaires, they were left to live in relative peace and allowed to participate in non-Indian society to a degree rarely seen elsewhere in California. The Yosemite's ability to adapt to their new world also made them inconspicuous to state officials, who had taken over Indian policy in California after federal efforts to develop a reservation system in the Central Valley failed in the early 1860s.[22]

Yosemite's Indian Wilderness

Despite the state's lack of concern, the presence of Indians in Yosemite proved a matter of considerable interest for many early visitors. The often patronizing affection that many tourists had for the Indians who lived in the valley, and the Yosemite's ability to reciprocate and even exploit these affections, went far toward ensuring they would remain in the area long after it became a national park. As Europeans and Americans had for the previous century and a half, early visitors continued to associate Indians with wilderness, and many were delighted to find them still living in Yosemite. A number of tourists happily recalled being entertained by their native and nonnative guides with accounts of Yosemite legends; still others commented excitedly about encounters with local Indians. The native settlement just outside the valley at Wawona became something of a tourist attraction in itself, and the "sweat house" there was an especially popular "object of curiosity."[23] Tourists would often visit the camp in the evenings to see how the inhabitants lived and at times dined with them in their dwellings.[24] In both Yosemite Valley and Wawona, the expertise of native hunters and fishers frequently received praise, and the daily chores associated with gathering, storing, and preparing acorns fascinated countless visitors.[25]

The association of Indians with wilderness was especially strong for early tourists, and one visitor in the 1850s even suggested that Yosemite be left entirely to the native residents. Unlike rapacious Americans, he observed, they showed their "love for the spot the 'Great Spirit' has made so lovely, and hallowed as the hunting ground of [their] forefathers."[26] After the creation of Yosemite Park in 1864, another tourist expressed similar sentiments in even more patronizing and romantic language. Thrilled that Yosemite was still home to "Indians, the simple children as of old," he wrote excitedly of "their bows, and arrows with flint heads; their food mostly acorns pounded in a rock hollowed out perhaps centuries ago for the same purpose; their furniture willow baskets; cooking by heating stones, and throwing them when heated into water; their faces tattooed and painted, and their enjoyments nothing above those of the animal." The government act to set aside a place still inhabited by these "simple children" gave him hope that "the time will never come when Art is sent here to improve Nature."[27]

The idea that Indians somehow complemented or completed a wilderness scene was also evident in the works of Yosemite's early landscape painters. While images of modern tourists in Yosemite could detract from the sublimity of the landscape, "picturesque" Indians or Indian-built structures further "naturalized" the scene and provided a human scale by which to emphasize the grandeur of the valley's cliffs and waterfalls.[28] The artist and writer Constance Fletcher Gordon Cumming, for instance, found Yosemite Indian encampments to be "filthy" and uninviting, but she could not resist placing them in the foreground of some of her paintings since they brought a "naturalness" and "blessed" touch of color to her art.[29]

James Hutchings, one of the valley's earliest and most avid promoters, clearly understood the tourist's fascination with Indians. In his many promotional writings about Yosemite in the 1870s and 1880s, he frequently called attention to the "Indian Camp, and its interesting people [as] . . . one of the many attractive features of Yosemite." For Hutchings, the native residents possessed "the principal customs, occupations, manner of living, habits of thought, traditions, legends, and systems of belief" not only of their own people and the surrounding tribes but also of "the California Indians generally." Consequently, the valley was an excellent place to see *real* Indians in their *natural* environment.[30] Though his comments reflected the romantic hyperbole of the time, in some respects Hutchings was right. The Yosemite probably constituted the largest native community in the central Sierra Nevada at this time, and their efforts to coexist with nonnative society actually preserved a high degree of cultural continuity and independence. Of course, they had adapted a number of their white neighbors' tools and customs, and the valley's roads, pastures, hotels, and campsites were anything but "natural," yet most early tourists simply applied a little imaginative effort to visually edit out such distractions.

Probably the most popular native occupations for early tourists was basketry, and many proclaimed Yosemite's basket weavers the finest in the world. The first recorded sale of a basket to a tourist in Yosemite occurred in 1869, but sales did not become commonplace until the 1890s. By that time, Miwok and Paiute women in and around Yosemite began manufacturing items expressly for sale to tourists. Their work soon became so famous that collectors and dealers regularly traveled thousands of miles to purchase baskets.[31] As Craig Bates and Martha Lee have observed, the Yosemite baskets were especially popular with tourists because they "brought to mind western, romantic, and primitive connotations." More than collectible items of merchandise, they allowed the purchaser to "sustain memories of their wilderness experiences."[32]

Baskets also represented an important means by which Yosemite Indian women could directly tap into the tourist trade and gain esteem in their own community. Basket making was a highly valued skill among the Yosemite, and though a woman could make more money as a laundress, the numbers and quality of baskets that a family possessed were a traditional sign of wealth and status within the community.[33] Consequently, a successful basket maker not only profited from the tourist trade but also utilized a skill that brought her respect from tourists, park officials, and other Indians. In doing so, she greatly enhanced her family's and her own standing within the larger Yosemite community.[34]

Lena Brown and Mary, Yosemite Valley ca. 1886. The daughter and granddaughter of important Sierra Miwok leaders, these two women held central roles in the Yosemite Indian community through the first decades of the twentieth century. They are pictured here in front of a summer *umucha*, a typical open-front dwelling still in use at the time. (Courtesy of the Yosemite National Park Research Library.)

Aside from basketry, native people found other means for profiting from the interest of early tourists. By the early 1870s, individuals would frequently entertain visitors outside their hotels and charge a penny for a brief dance or song. Larger "fandangos," as early Californians called them, might also have been held on occasion for the paid entertainment of tourists.[35] The growing popularity of Kodaks in the late 1880s made photographing Indians another important feature of the tourist experience. The Yosemite quickly recognized the marketability of their own "exotic naturalness," and several early tourists made special note of "a very cunning little papoose [who] smiled for a dime a smile."[36] Within a few decades, the price for a picture had risen considerably, and one popular basket weaver charged tourists a half dollar to photograph her with her wares.[37] In a 1904 book addressed to a growing interest in the Yosemite, Galen Clark admonished tourists not to expect the Indians "to pose for you for nothing [since] they are asked to do it hundreds of times every summer, and are entitled to payment for their trouble." He further advised his readers to "treat the Indians with courtesy and consideration, if you expect similar treatment from them."[38] By the turn of the century, native people had become an important part of the tourist experience, whether as laborers in the valley's growing service industry or as an authenticating aspect of the encounter with wilderness. Likewise, tourists had become an integral part of native people's lives; as one frequent visitor to the valley commented, a number of families were "in the habit of repairing yearly to . . .

Yosemite for the purpose of sharing in the double harvest—first of the tourists, later of the acorns."[39]

The presence of Indians in Yosemite during the last decades of the nineteenth century contrasts markedly with the policies of Indian removal implemented at Yellowstone in the 1880s. Established in 1872, only eight years after President Lincoln signed the Yosemite Park Act, Yellowstone is a near contemporary of Yosemite in the annals of wilderness preservation. The removal and exclusion of Indians from Yellowstone points up some significant differences in the evolution of these parks, however, and highlights the unique conditions that fostered the continuing development of Yosemite's Indian community. Because Yellowstone was created in Wyoming Territory, the issue of Indian removal from the national park was originally a federal prerogative. Consequently, park administrators could coordinate their efforts to exclude Indians from Yellowstone with officials in the Department of Interior, the Bureau of Indian Affairs, and the War Department. Yosemite, by contrast, was established within a state, and California officials retained sole responsibility for the valley's management until 1906. Like the state's management of Indian affairs, however, Sacramento took almost no interest in the administration of Yosemite. Even if state officials decided to exclude native peoples from Yosemite, their removal from the park would have been complicated by the fact that, after the demise of California's reservation system in the 1850s, there were no parcels of land to which they could be restricted. As a result, no policy ever developed regarding the removal or restriction of the Yosemite Indians, so long as the park remained under state control.

The different conditions surrounding the administration of each park certainly influenced the development or absence of a policy toward native residents, but the issue of their removal from park lands ultimately depended on the attitudes of park officials and tourists. Coming only a few years after George Armstrong Custer's debacle at Little Big Horn, the early exclusion of Indians from Yellowstone reflected a concern that they might frighten potential visitors away from the park. Unlike the tribes of the Rocky Mountain region, however, California Indians rarely marshaled a threatening resistance to the invasion of their homelands. Consequently, the presence of Indians in Yosemite Valley never became a matter of fearful concern among administrators or visitors. As one tourist observed in 1872, the Yosemite were altogether "mild" and "harmless," and wholly unlike the more dangerous tribes further east.[40]

Preservation and "Moral Rights"

By the 1890s, park officials at both Yosemite and Yellowstone began to share similar concerns about the presence of Indians within a nature preserve. In Yellowstone, Bannock hunting parties still frequented the park, and their presence was a matter of great consternation for park officials. Because the conflicts of the 1870s had already become a dim memory and the Bannock moved through only the most remote portions of the park, officials no longer worried that the presence of Indians might frighten visitors. Instead, their concerns reflected new ideas about Indians as both harmful to wilderness and potentially assimilable into American society. Yellowstone Superintendent Captain Moses Harris underscored this point when he argued not only that "marauding savages" threatened the wild flora and fauna in the park but also that Indians could never become "civilized" so long as they continued to frequent their former "wilderness haunts."[41]

Such ideas informed policies at Yosemite as well, and the establishment of Yosemite National Park in 1890, which then consisted of a large area surrounding the state-managed valley and Mariposa Big Tree Grove, brought new restrictions to the native community. The active enforcement of trespassing and hunting regulations, for instance, adversely affected those Indians who still hunted large and small game or gathered plants in the Yosemite high country. Unlike Yellowstone, however, native people still made up a significant portion of the park's labor force, and the idea that they somehow harmed wilderness did not lead to their outright exclusion from the more heavily developed valley. Furthermore, as Superintendent A. E. Wood noted in 1892, their long, unthreatening presence gave the Indians a "moral right" to remain in the state park. Wood also implied that removal would never really be necessary because the Yosemite were a "vanishing" tribe that would soon die out or assimilate into white society.[42]

Although Yosemite tourists and park officials generally had a more favorable attitude toward native peoples than did their counterparts in Yellowstone, a number of important early visitors complained about the presence of Indians in the park. In part because they did not match the "handsome and noble" Indians of popular fiction and art, the famous Unitarian minister Thomas Starr King visited Yosemite in 1860 and found the "lazy, good for nothing, Digger Indians" to be wholly incongruous with his notions of "pristine" nature. The fact that they gathered acorns from woodpecker stores only proved that Indians degraded the wilderness. Starr King felt that "many a Californian, if the question were up between the Diggers and the woodpeckers, would not hesitate in deciding the point of the 'moral value' in favor of the plundered birds" and seek to remove the Indians from Yosemite.[43]

Self-appointed "Friends of the Indian" such as Helen Hunt Jackson shared this disdain for the Yosemite. But for Jackson, wilderness also represented the depraved condition from which savages needed uplifting. Such "uplifting," incidentally, benefited the wilderness and, as Jackson noted during a trip to Yosemite in the 1870s, the presence of "filthy" Indians only detracted from the sublimity of the scenery. Furthermore, the inability of their "uncouth" minds to appreciate the beauty that surrounded them was an affront to the Creator and his works.[44] Like Jackson, John Muir found the Indians of the Yosemite region to be "mostly ugly, and some of them altogether hideous." Indeed, it "seemed [they had] no right place in the landscape," and Muir could not feel the "solemn calm" of wilderness when he was in their presence.[45] Starr King, Jackson, and Muir did not speak for most early visitors, but the longer the Yosemite persisted in the park and refused to vanish, the more such attitudes would drive park policy and eclipse any concerns about "moral rights."

The Yosemite, for their part, were not always happy with their non-Indian neighbors and the changes that had been wrought in the valley. In the late 1880s, a large group of Yosemite leaders sent a "Petition to the Senators and Representatives of the Congress," in which they complained of being "poorly-clad paupers and unwelcome guests, silently the objects of curiosity or contemptuous pity to the throngs of strangers who yearly gather in this our own land and heritage." They further noted that cattle and horses in the valley destroyed "all of the tender roots, berries and the few nuts that formed the[ir] sustenance." "The destruction of every means of support for ourselves and our families by the rapacious acts of whites," they continued, "will shortly result in the total exclusion of the remaining remnants of our tribes from this our beloved valley." In compensation for these damages to their homes and their way of life, they requested $1 million from the federal government "for the future support of ourselves and our descendants." In exchange, they promised to relinquish their "natural right and title to Yosemite Valley and our surrounding claims."[46]

None of the fifty-two men and women who placed their marks at the end of the document could have written it. Most likely, the author was the artist Charles D. Robinson because much of the wording is similar to complaints he brought before the California State Assembly during its investigation of the Yosemite Park Commission.[47] In the late 1880s, the commission had come under increasing criticism for its management of the park, and in response to these complaints the Assembly launched an investigation in February 1889. During public hearings, Robinson and others had criticized the commission's promotion of commercial development in the valley and its neglect of what they perceived to be its primary responsibility, the protection and preservation of Yosemite's natural environment. [48]These concerns were also included in the petition to Congress, but few if any of the Indian leaders would have troubled themselves with these political matters. Nevertheless, all of those who placed their marks at the bottom of the petition assented to its contents and certainly supported the author's intentions.

No one advanced the Indians' concerns at the state hearings, nor did they receive an answer from Washington in response to their petition. The hearings did much to damage the commission's reputation, however. In the aftermath, preservationists successfully petitioned the federal government not only to take over the management of Yosemite but also to considerably extend the park's boundaries. As noted before, the creation of Yosemite National Park in 1890 incorporated the high country surrounding the valley, thus protecting the area's flora and fauna as well as the streams that supplied Yosemite's magnificent waterfalls. The State of California held on to both Yosemite Valley and the Mariposa Grove for several more years, but these areas reverted back to the federal government and became part of the much larger national park in 1906.[49]

Federal administration of the areas surrounding the valley quickly became an active and proscriptive force in native peoples' lives. Management of the park by the U.S. Cavalry, which had taken over the care of Yellowstone as well, subjected them to all federal laws and park regulations. Before 1890, for instance, the Yosemite hunted deer throughout the Merced and Tuolumne River watersheds,

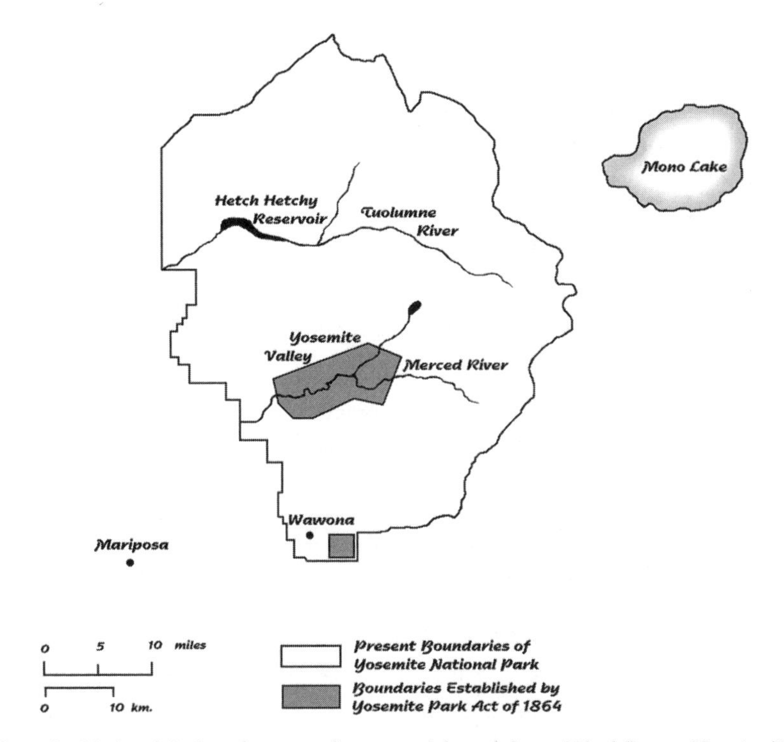

Yosemite National Park and surrounding area. Adapted from Alfred Runte, *Yosemite: The Embattled Wilderness* (Lincoln: University of Nebraska Press, 1990); and Craig D. Bates and Martha J. Lee, *Tradition and Innovation: A Basket History of the Indians of the Yosemite–Mono Lake Area* (Yosemite, Calif.: Yosemite Association, 1991).

but the cavalry severely restricted such activities within the boundaries of the new national park.[50] Hunting was absolutely prohibited, whether by Indians or by whites, and early superintendents aggressively sought to enforce the ban. In 1897, distressed that native hunters had killed a large number of deer the preceding fall, Acting Superintendent Alexander Rodgers insisted that "the interior department . . . take steps to prevent a recurrence of this conduct on the part of the Indians." Rodgers's recommendation was apparently heeded because later reports regularly noted that hunting within park boundaries no longer posed a problem.[51]

These new regulations reflect the zeal of military administration in the national parks, but they also demonstrate that late-nineteenth-century ideas about wilderness as uninhabited and pristine and about Indians as both vanishing and assimilable had begun to take hold in Yosemite. In many respects, the new restrictions placed on Yosemite Indian life mirrored the same mind-set that inspired the creation of Glacier National Park. As at Glacier, turn-of-the-century romanticism for the frontier inspired a sentimental interest in the Yosemite Indians that seemed to grow only stronger as native lifestyles "vanished" further into the past and as older, more "authentic" Indians died. As one tourist noted in 1913, the Yosemite lacked all "the picturesqueness which is so noticeable a feature of the red men in their native estate. . . . Nevertheless[,] the reason for their being in the Valley at all at this day, lends a certain romantic interest to their presence."[52]

Native people who did not look appropriately "Indian" presented a unique problem for park officials. On the one hand, they bolstered easy assumptions about vanishing or assimilating peoples, but on the other hand they disappointed tourists who wanted to see picturesque communities. Native people who attempted to practice older traditions, however, were somehow out of place because they had also adapted to new conditions and no longer seemed appropriately aboriginal. This apparent cultural disjunction had troubled tourists for some time, as well as the Army superintendents of the national park, but it apparently never concerned the men who managed the valley and Mariposa Big Tree Grove for the State of California. With the incorporation of these two areas into the national park in 1906, however, these concerns became the cornerstone of federal park policy toward the Yosemite Indians for decades to come.

As they took a more active interest in the valley's native community, national park officials also redoubled their efforts to restrict native use of the backcountry. The sentimental interests of tourists and the acknowledged "moral rights" of the Yosemite still precluded any efforts toward Indian outright removal, however. Instead, they inspired a series of administrative plans to incorporate the native community into official park promotions. In the summer of 1914, Acting Superintendent William Littebrandt urged the secretary of the Interior to bring the Bureau of Indian Affairs into a plan that would make the Indian village into "one of the features of the Valley, by attempting to reproduce a village or camp such as the Indians originally built."[53] The notion of constructing an "authentic" village for tourism was opposed by C. H. Asbury, a special agent for Indian affairs in the region, who strongly recommended against "establishing an Indian camp in the Valley, for exhibition purposes." As he noted in a letter to the commissioner of Indian Affairs, "The Indians . . . are there for the purpose of making their living at honest labor . . . and should be encouraged to make their own living, rather than become members of an aboriginal show."[54]

Park officials disagreed with Asbury's conclusions and continued to press their case. In an internal memorandum to park employees that was later forwarded to the commissioner of Indian Affairs, Littebrandt argued that a redesigned native village, if viewed with a "liberal" mind, "would not be detrimental to the interests of the Indians merely because it would [support] the interests of the park; the interests of both might be identical." "In other words," he continued, "an Indian is just as much a part of the scheme of things if he becomes a picturesque part of the landscape, as when engaged in some ugly and dirty employment; in both cases he is being used in the 'interest' of other parties, since few people employ them for benevolent purposes in any line."[55] Although this line of reasoning clearly advances one form of acknowledged exploitation over another, Littebrandt's tortured logic also reveals the close links between tourism and the presentation of past-tense Indian culture. Because his proposal coincided with plans to develop new park facilities near the site of the "old Indian camp," Littebrandt's argument also suggests that management of the park landscape would necessarily involve an effort to relocate the valley's native community to a more "appropriate" setting.[56]

Littebrandt had pitched his earlier appeal for a new village in terms of improving the housing conditions of the valley's native inhabitants. Calling attention to the community's general poverty, he "hope[d] that they may receive such assistance as the Government in fulfillment of its generous Indian policy may feel disposed to bestow upon them."[57] In a peculiar extension of the superintendent's own reasoning skills, the Indian Service declined to assist the Yosemite, who were "in no more need of aid, than [a] thousand others through the foot hills of California."[58] More particularly, the Yosemite Indians had never signed a treaty with the United States and thus had no official relationship with the federal government.[59] Consequently, Asbury had the same response for park officials that he gave an important Yosemite Indian leader named Francisco Georgely: if native residents wanted an improvement in their living conditions, as Georgely had petitioned the Indian Service, or park officials hoped to create a more "picturesque Indian camp," Asbury concluded they should both do so "at their expense."[60] The commissioner of Indian Affairs apparently agreed, and neither Littebrandt nor Georgely received any further response to their petitions. As later superintendents would soon learn, the Yosemite Indians were the exclusive "problem" of the newly established National Park Service, and one they would have to resolve on their own.

8

YOSEMITE INDIANS AND THE NATIONAL PARK IDEAL, 1916–1969

All fixed up! Ahwahnee too dirty bushy. . . . White men drive my people out—my Yosemite.

Totuya (1929)[1]

THE CREATION OF THE NATIONAL PARK SERVICE in 1916 fulfilled preservationists' long-held dreams for a strong federal commitment to the preservation and enhancement of all national parks. In many respects, this new branch of the Interior Department resulted from a six-year political battle against the City of San Francisco's plans to dam the Tuolumne River and convert the Hetch Hetchy Valley into a huge municipal reservoir. Because Hetch Hetchy was entirely within the bounds of Yosemite National Park and possessed scenic qualities that rivaled those of the more famous valley to the south, a coalition of public officials, civic groups, and national preservationist organizations joined with John Muir and others to protest what they saw as a fundamental violation of the national park and its boundaries. Their arguments failed to overcome the powerful thirst that would dam the Tuolumne River in 1913 and continued to drive the rapid development of the San Francisco Peninsula. Nevertheless, they did inspire the creation of a new government agency solely dedicated to the management and protection of national parks.[2]

The so-called Organic Act of 1916, which established the basic guidelines by which the National Park Service would manage its holdings, could not pull the plug on Hetch Hetchy and rescue the drowned valley. The act did promise to strengthen park boundaries, however, and declared their "fundamental purpose [as the conservation] of the scenery and the natural and historic objects and the wildlife therein." Still, the creation of the park service reflected more than the deep emotional concerns of people like Muir, who died shortly after the loss of Hetch Hetchy. Advocates for a strong park service believed that only broad popular support, based on wide use of the parks, would ensure the agency's continued strength and growth. Consequently, the Organic Act also mandated that the new government agency must "provide for the enjoyment" of visitors through the development of new accommodations.[3]

In practice, efforts to promote the parks as national "pleasuring grounds" far surpassed any concerns about trying to maintain them in what Secretary of the Interior Franklin Lane somewhat disdainfully called an "*absolutely* unimpaired" condition. In a letter to Director of the National Park Service Stephen T. Mather, Lane encouraged him to develop the parks as a new "national playground system" that should be made accessible to the public "by any means practicable." These included the construction of roads, trails, and buildings, and active cooperation with tourist bureaus, chambers of commerce, and automobile associations.[4] Moreover, as the Organic Act clearly stated, the director of the park service could also "dispose of timber" or "provide . . . for the destruction of such animals and of such plant life as may be detrimental to the use of any [park.]" If a tree or cluster of shrubbery blocked a certain view, for instance, it should be cut back or eliminated. Likewise, predators would have to be destroyed to increase the numbers of popular game animals like deer or mountain sheep that tourists expected to see in a national park.[5]

Instantly regarded as an expression of the guiding principles for the new park service, Lane's letter could not have found a more eager recipient than Stephen Mather. During his twelve-year tenure as director of the National Park Service, Mather followed an aggressive policy of park development that often led to now unthinkable proposals. At Yosemite, for instance, he championed the construction of a golf course in the valley, the building of a road around Nevada and Vernal Falls that would have connected the valley with Tuolumne Meadows, and even proposed that Yosemite make a bid to host the 1932 winter Olympics. Although none of these specific plans ever materialized, in part because they drew strong criticism from several quarters, all serve as important examples of the basic philosophy that would guide the management of Yosemite and other national parks at least until the mid-1930s.[6]

The Indian Field Days

If anything, the creation of the National Park Service only perpetuated the same struggle between preservationists and development interests that plagued the management of Yosemite in the 1880s. Like those earlier debates before the California State Assembly, this struggle would also carry over to the new agency's relationship with the Yosemite Indians. In the same year that the park service was established, Yosemite officials and concessionaires inaugurated the Indian Field

Days, a festivity designed to "revive and maintain [the] interest of Indians in their own games and industries, particularly basketry and bead work." The Field Days also encouraged visitation to Yosemite during the late summer, when waterfalls had either diminished to unspectacular trickles or dried up altogether. Any effort to represent or honor native culture, it seemed, must necessarily take place within concerns about attracting park visitors and providing for their enjoyment.[7]

Instead of promoting an interest in native "games and industries," park officials sharply circumscribed expressions of Yosemite culture. Event organizers encouraged Indians to conform to a generic representation of Great Plains culture, and the Field Days often degenerated into little more than an excuse for tourists and park officials to pose in buckskin and feathered headdress. Native people remained the central attraction of these events throughout the 1910s and 1920s, but only through their confirmation of popular white conceptions of how Indians were supposed to look and behave.[8] Basket judging and the sales of native crafts, for example, took place in front of crudely constructed canvas tepees. One year, in an attempt to lend some authenticity to the events, Don Tressider, president of the Yosemite Park and Curry Company, even looked into purchasing a wigwam from a group of Indians in Oregon. The traditional Miwok *umucha*, a conical structure made of long bark slabs, apparently reminded him too much of the Yosemite Indian village, which he regarded as an unpleasant eyesore that failed to satisfy expectations of native culture and life.[9]

Besides basketry and beadwork competitions, the 1925 Indian Field Days included a parade, rodeo events, an Indian Baby Show, and horse races featuring bareback riders "striped as Warriors." To encourage native participation in these events, park officials paid each man registered $1; every "squaw" appearing in "full Indian costume of buckskin dress, moccasin, and head decoration," garments wholly foreign to Sierra Miwok culture, received $2.50. The winners of "Best Indian Warrior costume" and "Best Indian Squaw costume" received $25 each.[10] Similar contests with similar incentives were a standard feature of all Field Days, and insofar as native people were encouraged to practice their "games and industries" at all, the park service and concessionaires expected them to fulfill popular conceptions of what Indians supposedly did.[11]

Along with promoting such stereotypical presentations of native culture, park officials strongly rebuked certain behavior as unacceptable. At the Indian Field Days of 1924, for instance, those attending the rededication of the Yosemite chapel heard a commotion from a group of Indians in the midst of a tug-of-war game a short distance away. A ranger rushed over from the chapel, ordered them to stop, and, because some had been excitedly betting on the contest, chastised them for gambling in a national park. A number of spokespeople for the Yosemite described the event in a letter addressed to the chief ranger: "The Indians were playing Tugo-war[;] the first game no one interfered, the second game Mr. Mather rushed in, and said no gambling in Yosemite National Park and ordered the Indians to leave this minute."[12] From subsequent correspondence among park officials it is not clear whether Director Mather, who participated in the chapel dedication, was the person who ordered the Indians to leave. Nevertheless, they certainly perceived

Indian Field Days, 1925. Leanna Tom, a Yosemite Indian woman born in the valley around 1850, holds here prize-winning basket at an Indian Field Days basketry contest. To her left are Mary Wilson and Alice James Wilson. Both are dressed in the type of "full Indian costume" that event organizers encouraged—and paid—all participants to wear. (Courtesy of the Yosemite National Park Research Library.)

the ranger's orders as representing the full authority of the National Park Service and resented the considerable attention park officials placed on this minor incident. Indeed, the tug-of-war game generated a surprisingly large body of correspondence among national park administrators and rangers, who eventually decided that the Indians would not be fined for gambling but must be further informed of park regulations and the consequences of ignoring them.

Along with gambling, park officials did not tolerate drinking or theft among native people or tourists. The penalties for Indians, however, were especially severe. In December 1925, Alvis Brown and Lawrence Beal, both twenty-one years old, were charged with theft and "sentenced to" a Bureau of Indian Affairs school in Salem, Oregon. A month later, fifteen-year-old Lawrence Dick received the same punishment for the same transgression and soon found himself almost six hundred miles from home and family.[13] Though attendance at government boarding schools was fairly typical for most Indian youths at this time, students frequently viewed their education as prolonged ostracization and punishment. Julia Parker, a Kashia Pomo woman who moved into the Yosemite Indian village after marrying Ralph Parker and joining his family there in the 1940s, met her husband at a Bureau of Indian Affairs school in Carson City, Nevada. "Boss Indians Around School," as she and her friends called it, was a place where native children were told to deny their heritage and trained to be "a person who was just a servant."[14]

Besides sending Indians out of Yosemite and placing them under the authority of another government agency, park officials also meted out their own punishments within the valley. In April 1926, for instance, park rangers arrested Virgil Brown for drunken driving and held him in the park jail for thirty days, an especially severe punishment for the time, and then banned him from the park. Always a favorite pastime with the Yosemite and many other American Indian groups, gambling was often an integral part of social gatherings. Nevertheless, the park service prohibited gambling, and after the tug-of-war incident rangers vigorously enforced this ban among the valley's native residents. At a 1928 Big Time, an annual summer celebration among the Yosemite and surrounding communities, park rangers arrested and fined Wesley Wilson for gambling with a "man and two Indian women." Neither the man, who was apparently white, nor the two women received even a lesser fine.[15]

Restrictions on Indians could also be accompanied by well-intentioned patronizing. Park administrators often acted as unofficial Indian agents and arranged for the health care of the valley's native residents. Partly to encourage participation in the Field Days, the park service also worked in conjunction with the California Bureau of Child Hygiene to provide a "well baby" checkup for participants in the Indian Baby Contest.[16] In 1930, when a seventy-two-year-old Yosemite man named Charlie Dick became too ill from tuberculosis for successful treatment at the valley clinic, Superintendent Charles Thomson arranged for his care in the town of Coulterville.[17] Although Dick paid for his own care, he apparently did not realize that he was doing so because Yosemite officials had long withheld part of his wages, without informing him, for just such a medical emergency.[18] In another instance involving money, however, Assistant Superintendent E. P. Leavitt helped Maggie Howard with a number of problems she had with the Bank of Italy in Merced.[19] Though these examples illustrate a sometimes benevolent interest in the welfare of the Yosemite, they were part of the omnipresent and intrusive role that park officials increasingly played in the Indian community.

In a strange and unsettling way, these efforts to control the valley's native population had strong parallels with other aspects of park management. Much as the Indian Field Days fit within a larger emphasis on making the national parks into places of recreation and entertainment, the restrictions placed on the Yosemite Indians also reflected the methods used to control the tourists' experience of the park's environment. The studied placement of scenic roadside overlooks, the cutting of timber to enhance certain views, and the tight management of animal populations were all designed to create what historian Richard Sellars has called "the scenic facade of nature, the principal basis for public enjoyment."[20] Whenever the behavior of native people infringed on the "facade" that park managers wished to create, their actions were sharply circumscribed.

Of course, if particular activities like the Indian Field Days contributed to the public's enjoyment of the park, then a native presence was strongly encouraged. As a general rule, however, park officials preferred to keep Indians outside the tourists' gaze. After the end of the last Field Days in 1929, for instance, they entertained a plan for encouraging a "thickening of undergrowth" near the Indian village to "segregate [it] from [the] public as desired."[21] Out of sight was not necessarily out of mind, however, and park officials seemed to view the entire native community as a potential problem that needed constant watch. Certain behaviors might be kept in check, much like a tree that threatened to block an especially photogenic vista, but some Indians, as young Lawrence Dick learned, could not be made to fit the park's management scheme. Like those bears that "misbehaved" in the park, any village residents who acted in a socially unacceptable manner would be banished from their homes in the valley.

Despite such encroachment into their lives, the Yosemite Indians successfully adapted to changing conditions in the park and, whenever possible, exploited them to their own advantage. At the turn of the century, for instance, they lived in six small encampments from spring through fall but gradually merged into one larger village. This change strengthened the community as a whole and better accommodated the Indians to Yosemite's ever-increasing tourist development. As Lowell Bean and Sylvia Brakke Vane have noted, such an important social change took place along traditional lines, and community leaders continued "to maintain older religious and political structures" as they worked to bring native life into accord with new developments. Such qualities proved essential through the 1910s and 1920s, and the valley's native community managed to ignore a certain level of outside intrusion as they successfully adapted to new developments.[22]

While the Yosemite affirmed long-established social structures, they apparently had no qualms about participating in the cultural novelties of the Indian Field Days. Local basket weavers looked forward to matching their skills with those of neighboring women, and the festivities drew a large number of customers for their wares. Likewise, the rodeo provided a public arena for Yosemite men to test their riding skills—and their luck at betting—against other Indians from around California and Nevada. In short, they participated in the Field Days because they enjoyed the events and derived certain benefits. Nevertheless, the event did not become an especially important part of native life or supplant the traditional Big Times, which brought large numbers of Indian visitors into the Yosemite community for days and nights of feasting and dancing.[23] Perhaps more than anything else, the continuation of the Big Times through the first decades of the twentieth century illustrates the vitality of native traditions and the paramount importance of the Yosemite Valley for the region's larger Indian community.

Defining the Indian Problem

The strength of Yosemite's native community and its ability to remain in the valley met the greatest challenge in the late 1920s, when a new park master plan called for removal of the Indian village. Because the National Park Service wanted to build a medical clinic and store on the site, Superintendent Washington Lewis proposed moving the park's native residents to another location within the valley. Although Lewis had developed this plan without consulting the Yosemite themselves and he always found their village to be "more or less a nuisance," he did not entertain any notions of outright expulsion. The Field Days proved that visitors enjoyed seeing Indians in the park, and several years of these successful events had fostered a certain level of appreciation toward the valley's native residents. Recognizing their popularity among tourists, Lewis instead echoed the sentiments of his predecessors in the 1910s and proposed the development of a new village in "an Indian character design . . . thereby making . . . [it] a very presentable thing." Such a plan would not only satisfy the expectations of tourists but also promised to quiet a string of recent complaints from visitors about the unsightly poverty of many native houses.[24]

What such a design entailed was not altogether clear, but it certainly did not include improvements that the Yosemite might propose. While redesigning his old house in the village, Harry Johnson learned from park officials that he would have to cease construction because his additions were "too conspicuous from the road . . . and lacked the proper architectural lines." Johnson's house apparently did not look "Indian" enough to the administrators, or it simply clashed with the master plan's requirement that all new construction should reflect a certain "harmony with the landscape." Of course, it never bothered these park officials that the road near Johnson's house could also be seen from the road, nor did they consider how this road might or might not harmonize with the landscape. In either case, it obviously did not matter that Johnson's improvements grew out of his own feelings about what constituted an appropriate dwelling in the Indian village.[25]

Superintendent Lewis hoped the Bureau of Indian Affairs would help finance the proposed new Indian village and encouraged its contribution to the village's planning and implementation. Unfortunately for Lewis, the response of L. A. Dorrington, superintendent of Indian Affairs in Sacramento, was even colder than that given park officials some thirteen years earlier. Because the people who lived within Yosemite National Park had not signed treaties with the United States, Dorrington reminded Lewis, the Indian Service could not directly aid the development of the new village or contribute to the support of the Yosemite Indians.[26] In an apparent effort to soften this rejection and stave off any further requests, Dorrington did confide to the commissioner of Indian Affairs that he at least expected to take certain salutary measures that would make park officials "think and feel the problem is theirs and that we are helping solve it."[27] Despite the rebuff and perhaps believing that some support from the Indian Service might still be forthcoming, Lewis moved ahead with Yosemite's master plan and set out to determine how best to incorporate native people into the proposed improvements to park facilities.

As Lewis had informed Dorrington and others, the park service felt it necessary to limit the number of Indians living in the park to those individuals of "the original Yosemite band or their descendants."[28] After conducting the first in a series of Indian village censuses to determine who could remain in the valley, park officials apparently decided to extend these criteria to a slightly larger group of people. This may have occurred because Lewis and his successors tended, on the one hand, to favor a number of individuals who might not qualify as "true" Yosemite Indians—namely, popular basket weavers like Maggie Howard and Lucy Telles, who were both of mixed Mono-Paiute and Sierra Miwok ancestry. On the other hand, park officials had expressed a strong resentment toward some residents like Virgil and Alvis Brown, who were both descendants of important Ahwahneechee leaders. By the spring of 1928, then, the park service qualified their criteria for Indian residency, ruling that those individuals presently living in the village could remain only if they had an established "right to do so, either through being natives of Yosemite Valley or because of their long residence [there]."[29]

Much as they had qualified the basis on which native residence would continue, park officials soon equivocated on their definition of a "right" to live in the valley. By the summer of 1929, the issue had been thoroughly studied, and Lewis's successor, Superintendent Charles Thomson, met with the Indians in the village to "impress upon them in a proper way, that their residence [in the valley was] a privilege, and not a vested right; [and] that this privilege [was] dependent upon proper deportment."[30] He also told them that certain people, namely "the Yosemite Indians . . . and the Mono and other Indians who [had been in the park] for years and years[,] . . . had a 'moral right' to remain in the valley." Nevertheless, he warned that "should it prove to be in the best interests of the Government to build houses and assign them, it will give [park officials] absolute control of the Indian Village."[31]

For Thomson, the issue of control was paramount. "If anyone was constantly breaking a regulation," he told the assembled Indians, "did not want to work reasonably steady, cannot get along with his neighbors, or in any way prove to be a poor member of the Village . . . he would have to go away and give up his house." Furthermore, anyone who could not find work in the park during the fall and winter months would have to leave as well. As Thomson well knew, almost no one, white or Indian, worked in the valley during these months. Hence, Thomson's "absolute control of the Indian Village" did not simply mean a severe regulation of Indian life and a dismissal of the "moral right" to remain in the valley; it implied the possibility of outright eviction for the entire native population.[32]

Not surprisingly, such talk had a considerable effect on the Yosemite Indian community. Almost immediately, they turned for assistance to the Indian Board of Cooperation, a nonprofit legal organization in San Francisco.[33] The best way to resist possible eviction from the park seemed to lie in a strategy of refusing to move to a new village site. Moving would forfeit any inherent claims they might have on the present village area, and paying rent implied that continued residence in the valley was dependent on where and how the park service chose for them to live. Although this tactic had little chance of legal success, it received the personal support of the Indian Board of Cooperation's executive director, Frederick G. Collet. These developments created quite a stir among park officials, who referred to Collet as the leader of an indeterminate number of meddling "outside agitators." By December 1929, Thomson was "fed up" and resolved that he "must go gunning for [Collet]" before matters got out of control. While no one actually threatened Collet physically, park service officials did ask the U.S. Attorney General to investigate him—which apparently helped to quiet matters considerably.[34]

Although the Yosemite Indians never sought legal redress against the park service and government investigators seemed to have been successful in their efforts to undermine Collet and the Indian Board of Cooperation, Superintendent Thomson realized that he would need to tread more softly in his dealings with the valley's native residents. He still viewed the Yosemite Indians as less than "desirable citizens of any community," as he noted in a "Special Report on the Indian Situation" to National Park Service Director Horace M. Albright, and felt "they should have long since been banished from the Park." According to Thomson, their "ejection" would bring a number of great benefits: it "would ease administration slightly; would eliminate the eyesore of the Indian village . . . and, following the elimination of private land holdings [on the western perimeter of the park], would remove the final influence operating against a *pure status* for Yosemite." Echoing the sentiments of earlier Indian reformers, Thomson believed that removal would also benefit the Yosemite Indians because it would "tend to break them up as a racial unit and, in time, to diffuse their blood with the great American mass."[35]

Despite all these advantages, the superintendent still had to recommend against a concerted effort to remove the Yosemite from the park. "While their ejection might be the simple and easy solution," Thomson ultimately declared that he was "opposed to it." As the recent experience with Collet had proven, such a policy would raise a "storm of criticism [from the Indians and their allies] . . . that could hardly be withstood." Nevertheless, he was under considerable pressure to develop a solution to a problem that he felt could "not be tolerated much longer."[36] The park master plan had received final approval from several government agencies, and construction on the site of the Indian village was imminent. With short-term needs and long-term goals in mind, Thomson proposed a middle course that would give park officials unprecedented control of Yosemite's native community and, over time, achieve the full removal of Indians from the park through a process so gradual that it would not draw any adverse publicity.

Toward a Final Solution

Thomson's report became the definitive statement on park policy toward the Yosemite Indians and received enthusiastic support from both Albright and the Yosemite Board of Expert Advisors, a nongovernmental group established to advise Yosemite administrators on matters of policy and development. Although Thomson exhibited considerable disdain for the park's native residents, he believed their presence in the valley imposed an "obligation upon those charged with the handling of backward peoples." Moreover, their "historical association with Yosemite makes them very significant to the Park; to drive them out would result in an ethnological loss comparable to the loss . . . [that] our deer would mean to our fauna exhibit." Because some native residents were popular with visitors, "especially Easterners," Thomson also agreed with an advisory board recommendation for a native exhibit "done in the aboriginal style, with one or two Indian families resident, during the summer garbed in native dress, carrying on the pursuits of their forebears."[37]

Thomson's reference to the fauna exhibit is telling. Like the tame animals near the visitor's center, the display of past-tense Indians behind the new Yosemite Museum would certainly fit the park service's goal of presenting a "scenic facade of nature." In doing so, the proposed exhibit would also mark a sharp turn away from the more commercial qualities of the Indian Field Days. Not surprisingly, the same Board of Expert Advisors that encouraged the development of the "aboriginal style" presentation also roundly criticized the Field Days as "a white man's [entertainment], in which some part is taken by Indians to whose Yosemite forebears such things are wholly unknown." Of course, the new goal was not to make native people less entertaining or interesting but to present them in a more "authentic" manner.[38] Thomson fervently agreed and hoped the relocation of the new Indian village to a more secluded location in the park would also prevent its residents from maintaining their "tendencies toward professionalizing—fortune telling, fake Indian dances, etc. for fees."[39]

As they created a program for dealing with Yosemite's native community, park officials were also guided by a newly developing preservationist ethic. Beginning in the mid-1920s, biologists like Joseph Grinnell at the University of California and George M. Wright of the National Park Service sharply criticized any policy that placed the development of roads and hotels above ecological concerns. Focusing on wildlife management, which mainly consisted of predator reduction and feeding programs, Wright and two colleagues began a study in 1930 that advocated the restoration of park environments to their "pristine state."[40] Published three years later as *Fauna of the National Parks of the United States*, the study represented the first serious effort to move the park service away from the development of tourist amenities and toward a focus on the Organic Act's stipulation that "scenery," "natural objects," and "wildlife" should remain "unimpaired."[41]

Much of the criticism about excessive development in the national parks focused on Yosemite, and Superintendent Thomson often struggled to respond to these new arguments as he tried to fulfill an obligation to improve visitor facilities and increase public use of the park. Wright and his colleagues sympathized with the view that national parks existed mainly for public use, but they still believed that visitation should be better "reconciled" to existing biological relationships.[42] Thomson ridiculed such ideas and proclaimed in a July 1931 memorandum to Director Albright that "balances of nature or other hypothetical or similar theories" would not deter him from providing for the needs and expectations of park visitors.[43] Despite such protests to the contrary, Thomson did share some of his critics' concerns and, as noted above in his "Special Report on the Indian Situation," he believed that one of the park service's highest priorities should be the achievement of "a pure status for Yosemite."

For Thomson, the purity of Yosemite required more than a tightly managed presentation of waterfalls, forests, granite cliffs, and docile game animals. At the very least, it involved the extension of the park's western boundaries over several remote tracts of unlogged forest land.[44] This concern for an area that lay outside the purview of most visitation might seem ironic, but such apparent contradictions were part and parcel of the park service itself in the 1930s. Indeed, Thomson embodied many of the competing interests that both defined and undermined what historian Alfred Runte has called a "groping for awareness" of the need to preserve Yosemite as a representation of "original American wilderness."[45] Though only partly formed, it was this last concern that largely inspired Thomson's program for gradual removal of the Yosemite Indians. The banning of native hunting and the suppression of Indian-caused fires in the late nineteenth century had already gone a long way toward making both the valley and the surrounding high country into the type of well-wooded, game-rich landscapes that park officials and tourists preferred. The real problem, it seemed, now derived from the simple fact of residence.[46]

Of course, as Thomson informed Albright in his 1930 "Special Report," the "pure status of Yosemite" could not be achieved overnight. Nevertheless, construction of the new Indian village would give the park service tremendous leverage over the Indian population within the valley; "the Superintendent could prevent the influx of outside Indians and, by the device of cancellation of lease of those abusing the privilege of residence, he could maintain a discipline now impossible." Furthermore, Indians would have to pay rent, and those who fell delinquent in their payments or were absent from their homes for too long would forfeit their residences in the valley. Those gainfully employed by either the park service or one of the concessionaires could remain in the new Indian village, but all were to be retireable employees. And once retired, they had no right to remain in the valley—moral or otherwise. Ultimately, the native presence in the valley would cease to be a problem because it would eventually take care of itself through a process of attrition.[47]

The park service began construction of the new village in 1931, and six cabins were finished by mid-November. Mindful of native protests and fears about the loss of their ancient village site, Thomson "kept entirely away from the Indians until [after construction was finished], keeping them in suspense as to our plans." "As good luck would have it," he informed Director Albright, "their completion coincided almost exactly with the onset of [a] bad storm." Because "a foot of snow and very cold weather made the cabins even more attractive . . . [he] called a general meeting of the Indians" and informed them of their imminent move. Though three Yosemite leaders protested in a "suspicious and hard-boiled" manner, the families selected to live in the new cabins quickly gathered their belongings the following morning and "moved in very fast as if they feared [that park officials] might change [their] minds."[48]

Although most members of the Yosemite Indian community would remain in their old homes for another few years, the last residents finally moved into the new housing units in 1935. When completed, the new site contained twelve cabins for a permanent population of sixty-six individuals. The structures were tiny, only 429 square feet in size, and housed as many as six to eight family members.[49] The addition of three more buildings slightly alleviated cramped conditions, and by 1938 these fifteen cabins housed a total of fifty-seven people.[50] The small size of the new village was designed in part to prevent the "riffraff or the Indian population of the surrounding country" from "swarm[ing]" into the valley for work and residence.[51] More significantly, as Thomson had implied in his special report, the limited amount of living space would also let park officials choose who could remain in the park and who must leave.

Ostensibly reserved for those with the strongest "moral right" to reside in the valley, other criteria tended to be more important when designating who could live in the park. Not surprisingly, park officials did not allow Henry Hogan to move into the new village, though all regarded him as a "true" Yosemite, on account of his previous trips outside the valley to buy liquor. Authorities also denied Jim Rust a place in the new village because he "had no connection with the valley . . . beyond that of an ordinary laborer." Apparently, the relative unimportance of Rust's work in Yosemite was more significant than his being the great-grandson of Tenaya, the Ahwahneechee leader at the time of Yosemite's "discovery" in 1851.[52] As Thomson told the Indians during a meeting at the new village in the fall of 1931, their continued residence in the valley would depend less on ancestry and more on "usefulness to the community; length of service working in Yosemite; ability to support themselves; [and] number of children."[53]

New Indian village, 1933. In this official park service photograph, some residents of the Yosemite Indian Village stand in front of Harry Johnson's new home. Johnson's eldest son, Jay, would become the last Yosemite Indian to live in the park.

With residence so closely tied to employment, native protests against the park service largely concerned the availability of jobs. David Parker spoke for many when he wrote Commissioner of Indian Affairs John Collier in May 1933: "The working condition of the Indians in Yosemite . . . is not very good [and the] work we were offered is not carried out as promised. . . . The Indians are removed from their regular camping ground to different place, [and] . . . now we have to pay for the house and light bill . . . [w]hich some of us cannot afford to pay . . . until we get job from [park officials]." "I don't see why we are not getting any chance," he added, "we are dependable as any men." Because the very existence of the Yosemite Indian community was now made vulnerable to unemployment, Parker's complaints went far beyond a simple concern about falling behind in his bill payments. As he tried to make plain to Commissioner Collier, the security of the Yosemite Indian community depended on the strength of their connection to the valley.[54]

Though it still remained an important place for annual celebrations and provided a focus for native life in the region, eight decades of adaptation, accommodation, and subtle resistance had certainly transformed the Indian community's relationship to Yosemite. The valley itself had also changed: acorns, tubers, seeds, grasses, and game animals were less plentiful, and in some instances locally extinct; roads had replaced trails; trees and bushes filled the grassy meadows that had once been tended by centuries of careful burning; and park service buildings and tourist facilities clustered over abandoned village sites. Both people and place were profoundly altered, but their connection was no less diminished. If anything, the Yosemite had developed a new and perhaps deeper appreciation for their home. Nothing better illustrates this attachment to place and the fear that park officials were trying to destroy the Yosemite Indian community than a prophecy by Bridgeport Tom in the early 1930s. As his grandson Jay Johnson relates the story, Tom warned that the great cliffs of granite would collapse into the valley when the last of his people left. The end of the Yosemite Indians, then, would mean the end of the valley.[55]

At a certain level, park officials had recognized this continuing attachment to place in what they called the Indian community's moral right to remain in the valley. They conveyed this sentiment to Commissioner Collier when he inquired about the park service's employment of Indians but also defended their decision not to provide full employment for native residents of the valley. Because Parker had raised a fundamental question about park policy and the viability of the native community, park officials defended their actions by attacking his credentials as a true Yosemite Indian. Although he was born in Yosemite, where most of his family continued to live, and descended from people born in the valley before 1850, Parker was not considered a "local" because his most recent residence had been near Mono Lake. Consequently, any employment he obtained in the valley should be viewed as a favor, not a right. The explanation apparently satisfied the commissioner of Indian Affairs, and he informed Parker that any complaints against the park service were entirely unjustified.[56]

Of course, Parker's family viewed matters differently, and his inability to secure regular work signaled a threat to the entire community. If a recognized member could not return, then any temporary departure from the valley might prove permanent. Movement in and out of the area, as well as incorporation of individuals from neighboring groups, was a vital dynamic that had long shaped Yosemite Indian life. Now the park service had defined people like David Parker as part of a "large miscellany of Indians living in surrounding counties who naturally would like to be included into the advantageous status of our Yosemite group of Indians." Parker's complaint was shortly followed by another, but park officials apparently saw no reason to follow up on the matter. The park service's control of housing and employment in the new Indian village, as well as the recent vote of confidence from the commissioner of Indian Affairs, made their position unassailable.[57]

While the likes of Parker might be put off, Thomson's efforts to create a "pure status" for Yosemite National Park did not take effect as rapidly as he might have liked. As he noted in a letter to the director of the National Park Service in July 1933, the Yosemite Indians provided "a reservoir of almost efficient labor upon which [the park service and concessionaires could] draw," and rapid attrition of workers would have been counterproductive.[58] A certain balance had to be struck between the labor needs of the park, the desire to eliminate the native population altogether, and a fear that any sharp drop in the number of people residing in the Indian village might be construed as forced removal. Whether by design or simple prejudice, park officials found their solution in a policy of casual neglect. Throughout the 1930s, they regularly failed to assist with the maintenance of the new village, even as rents increased, and continually ignored earlier promises to give Indians first consideration for park employment. While part of an ongoing effort to prevent an "influx of other Indians [moving] into . . . favorable living conditions" in the valley, ignoring the concerns of the park's native community fit nicely within Thomson's original plan. As the superintendent had informed a group of U.S. senators who visited the park in the summer of 1932, "he did not want to encourage permanent residence in the park" but intended the condition of the village to foster a "tendency . . . to drift away" from the valley.[59]

As employment in the park became more difficult to obtain, individuals and families moved out of the valley to adjacent areas in Mariposa County. Despite new births within the Yosemite Indian community, the population of the village had been halved by the early 1940s, and many of the remaining residents were slated for retirement in the coming years.[60] As Jay Johnson recalls his years growing up in the valley, the people who continued to live in the Indian village deeply felt and resented the control that park officials wielded over their lives. Even those secure in their employment and residence had to contend with a growing list of regulations over personal conduct, the appearance of their homes, and the care of their children.[61]

Conditions in the valley only worsened for the Yosemite Indian community in the years following World War II. Park officials nearly tripled the rent in 1947 yet declined to make a commensurate increase in the services they provided to the Indian village. After a series of complaints from residents forced some concessions from the park service, relations between the two groups became increasingly antagonistic. Whether the result of protests from the native community or the product of increased surveillance, park officials recorded large increases in the number of Indians cited for violating park regulations. A supposed inability to respect authority seemed only part of the problem, and the Yosemite were also accused of not being "real" Indians. Besides acquiring many of the accouterments of modern society, the Yosemite's "blood-line" had become "very thin" through intermarriage. Consequently, bad behavior and corrupted blood meant the Yosemite Indians no longer possessed any moral right to live in the valley. As historian Stella Mancillas has noted, these conclusions would soon inspire the park service to take steps that would accelerate the demise of the Indian village.[62]

Conflict and hardship also tended to strengthen the Yosemite Indian community, and no one had any desire to leave the village. In fact, a number of individuals who served in the armed forces or took jobs in the burgeoning defense industries returned to Yosemite in the years after the war. Marriages and births also brought new members into the community, and by 1953 the year-round population of the village had climbed to forty. The valley also remained a focus of Indian life in the region, and residents continued to host "Feeds" and "Gatherings" on a regular basis. While a new generation learned songs, dances, and stories at these events, Yosemite elders also imparted a deep respect for their homes in the valley. Tensions between the Indian community and the park service only heightened this connection to Yosemite, and native elders encouraged their people to view residence in the valley as a cultural necessity.[63]

The small increase in the number of people living in the Indian village, and the fact that almost half were under the age of twenty-one, caused park officials to fear that Yosemite's "Indian problem" would continue to escalate indefinitely unless strict measures were adopted. In the summer of 1953, the park service developed the Yosemite Indian Village Housing Policy, which stipulated that only permanent government employees could remain with their families in the village. Those who did not meet this single qualification received notice to leave the valley in four weeks. Once vacated, their cabins were destroyed to prevent any members of the growing families that still remained in the valley from taking up residence.[64]

Superintendent Charles Thomson had once feared a public outcry against wholesale eviction, but park officials in the 1950s could view their actions in light of new developments in federal Indian policy. Much like the government's effort to "terminate" its relations with Indian tribes in the 1950s, the park service now argued that Indian removal from Yosemite would prove a blessing to the valley's native inhabitants: residence in the national park had sustained the Yosemite Indian community, but in the process it retarded an individual's ability to join the mainstream of American society. They expected that the Yosemite Indians, forced to live without the watchful direction of the National Park Service, would shed their collective identity and learn to fend for themselves. As Congress declared to Indian tribes across the country, park officials believed their new plans would allow each Yosemite Indian to finally enjoy "the same privileges and responsibilities as are applicable to other citizens of the United States, to end their status as wards of the United States, and to grant them all of the rights and prerogatives pertaining to American citizenship."[65]

Park officials rightly gauged public opinion, and implementation of the new policy elicited no appeals on behalf of the valley's native inhabitants. Unable to marshal an effective resistance on its own, the residents of the Yosemite Indian community were forced to comply with a government agency that now seemed to possess almost total control over their lives. Whenever a head of household died or retired, the park service never failed to issue an order to vacate the park. As each successive family left its home in the valley, the remaining residents of the Indian village could only watch as park employees destroyed or removed each newly vacated cabin. By 1969, only a few structures remained, and the last residents were relocated to a government housing area for park employees. Abandoned and dilapidated, the Indian village soon vanished in the flames of a fire-fighting practice session.[66]

In the years following the fiery destruction of the Indian village, the park service managed to erase almost all signs of habitation. A few traces of the Indian village still remain, however: a fire hydrant that stood near the common garage seems mysteriously out of place in the middle of a campground popular with climbers; gentle concaves and small holes made from the preparation of acorns adorn some unnoticed and out-of-the-way pounding areas; a number of cultivated plants have escaped efforts to return the area to its "natural state."[67] Small testimonials to almost forty years of habitation, these signs betray a longer history of adaptation and persistence that continues to shape the Indian people of Yosemite. Shortly after the final demise of the village, the Yosemite Indians reorganized themselves as the American Indian Council of Mariposa County (AICMC). Dedicated to strengthening older cultural practices and establishing official relations with the federal government, the AICMC has reinforced the connections between old residents of the valley with other native groups in the area. Together, they have asserted a cultural claim to Yosemite National Park that has allowed native people to regain some access to the valley and its resources.[68]

No longer residents in the national park, the Yosemite still have a close connection with their ancestral home, and many frequent the valley to gather acorns, celebrate the annual Big Times, and maintain traditional religious practices. The tremendous amount of tourist development in the park has compromised much of Yosemite's environment, however, and recent park service efforts to repair the damage have occasionally resulted in tighter restrictions on native utilization of park resources. Though Indian uses did not produce the current problems in the park, intermittent crackdowns on the gathering of certain plant resources has only added to the tensions that have long characterized the park service's relationship with the Yosemite Indians.[69]

Almost seventy years ago, Superintendent Thomson felt that the government had "solved a perplexing problem and would have no other task with [the Yosemite] except to prevent the influx of other Indians into these favorable living conditions." By establishing a plan through which the Yosemite would eventually be forced to leave the valley, and by segregating those who remained from more commonly visited areas of the park, Thomson achieved a solution to an issue that had bothered officials since the establishment of Yosemite National Park in 1890. The subsequent construction of a sanitized ethnological exhibit depicting pre–gold rush Indian culture further restricted the Yosemite Indians' visible presence in the valley and effectively contributed to a historical fiction still maintained by the National Park Service in its literature on Yosemite and most other national parks: Indians were the first "visitors" to park areas, who, for a variety of reasons, decided not to visit these lands sometime in the distant past, and, at least in the case of Yosemite, "real" Indians ceased to be a viable presence in the area long before the establishment of the national park.[70]

With the Indian "problem" solved and Yosemite no longer an anomaly in the national park system, such fictions have become further embedded in popular conceptions of national parks and wilderness. Americans look at an Ansel Adams photograph of Yosemite and see more than a national symbol. They see an image of a priori wilderness, an empty, uninhabited, primordial landscape that has been preserved in the state that God first intended it to be. Ironically, when Adams took his most famous photographs a sizable native community still lived in Yosemite—the descendants of the same people whose habitation of the valley in the mid-nineteenth century qualified Yosemite as a true wilderness in the minds of many Americans. What Adams's photographs obscure and what tourists, government officials, and environmentalists fail to remember is that uninhabited landscapes had to be created.

If Yosemite National Park teaches us anything, it is that scenes of great permanence are fraught with historical change. With every change, however, it seems there is always an ending and a new beginning. In late December 1996, the last Yosemite Indian to reside in the national park left his birthplace for a new home in the community of Mariposa. Jay Johnson, the eldest son of Harry Johnson and the grandson of Bridgeport Tom, had retired the previous July from his position as a forester with the National Park Service. In accordance with the Yosemite Indian Village Housing Policy of 1953, he and his family had to leave their home by the end of the year. On New Year's Day, his grandfather's old prophecy seemed to come true: a huge storm roared through the central Sierra Nevada, the raging Merced River tore through park structures, and huge rocks fell thousands of feet before smashing onto the flooded valley floor.[71]

The storms of January 1997 may have been only a warning. Johnson, his family, and members of the AICMC still regard Yosemite as their home, and all have struggled to maintain a connection to the valley. Consequently, they place great importance on a number of agreements with the park service that allow them to at least continue their ceremonial use of the park area. Most significantly, a recent compact between the National Park Service and the AICMC to convert the last village site into an Indian cultural center promises to greatly strengthen the native presence in Yosemite. Unlike the historical displays behind the Yosemite Museum, the new center would celebrate and foster the continuing vitality of the Yosemite Indian community. Moreover, the area will provide a permanent space for ceremonies and social gatherings that participants can choose to close to outside visitors. According to Johnson, the practice of certain rituals has long "kept things in balance," and the new cultural center will help to maintain the connection between his people and the place that has long sustained them. But Johnson warns that if access to the old village site is ever denied and the ceremonies could not take place there or elsewhere in the valley, "then watch out" for the true realization of Bridgeport Tom's prophecy.[72]

CONCLUSION

Exceptions and the Rule

JOHN MUIR ONCE DECLARED THAT true lovers of wilderness enjoy a "close and confiding union with [Nature]."[1] Having defined, created, and preserved the object of these affections, Muir and his friends could certainly lay a special claim to America's uninhabited wilderness parks. It is no great trick to love one's own creation, however, and scholars have recently begun to recognize a certain degree of narcissism in American conceptions of wilderness. For those native peoples who found themselves excluded from national park areas in the late nineteenth and early twentieth centuries, the cultural construction of wilderness was already old history. As Luther Standing Bear observed in the early 1930s, "Only to the white man was nature a 'wilderness,' and only to him was the land 'infested' with 'wild' animals and 'savage' people. To us it was tame." Likewise, a contemporary of Standing Bear's, Iktomi Lila Sica, characterized park service claims that Indians had not used preserved wilderness areas in the past as "ridiculous propaganda."[2] Long before Luther Standing Bear or Iktomi Lila Sica reached young adulthood, Shoshone, Bannock, and Crow people clearly understood the exclusive nature of wilderness and its appreciation, as would the Blackfeet and Yosemite a short while later. In time, the mostly unwritten experiences and resentments of these peoples would inspire later generations to challenge their continued exclusion from national park lands.

By the 1930s, the object of John Muir's affections had largely become a reality. Yet the exclusion of native peoples from national parks did not represent the end of Indian efforts to regain access to their former homelands. If anything, the 1930s marked the beginning of several new attempts to open up national park areas for traditional uses. Under Commissioner of Indian Affairs John Collier, fifty years of forced assimilation programs were replaced with new federal policies that supported a certain degree of cultural and political autonomy for many tribes. While no one within the Indian Service directly supported native claims to the national parks, Collier's Indian New Deal did foster a level of tribal activism that made it difficult for the park service to "preserve" more wilderness areas at the expense of Indian communities.[3] The example of the Blackfeet and Glacier National Park clearly demonstrates these new conditions, but similar developments took place at Canyon de Chelly National Monument and in the national forest reserves of the Pacific Northwest.[4]

One of the most remarkable proposals for national park management in the 1930s came from Iktomi Lila Sica, who called for the creation of several national "Indian-wild life *sanctuar[ies]*." Sounding like George Catlin before him and obviously sharing the concerns of Black Elk, he hoped to see Badlands National Monument, the Pine Ridge and Rosebud reservations, Black Hills National Forest, and Wind Cave National Park combined with other state and federal lands. Together, these would be "reestablished as a WILDERNESS AREA . . . FOR INDIANS AND WILD LIFE, and, for Indians and visitors, a scientifically zoned sporting, recreational, health and scientific area accompanied by conservation."[5] Such an area would provide income for Indian tribes and allow unused or "wasted" wilderness areas to be utilized in a nonexploitive manner. Besides South Dakota, Iktomi Lila Sica proposed similar combinations of Indian reservations and public lands around Yellowstone and Glacier national parks. These "inter-reservations" would not only provide important cultural use areas but also support several campuses of a national "Indian University" where native and nonnative students could study traditional and modern land management techniques.[6]

Iktomi Lila Sica's program for "American Indian regeneration" partly smacked of George Catlin's romanticism and certainly shared much of the New Deal's optimism for grandiose public lands projects. Nevertheless, it stemmed from a belief that traditional land use practices must be maintained to guarantee the survival and future health of native societies and the environments on which they historically depended. In light of the recent events at Glacier and Yosemite national parks, however, these ideas had no chance of affecting Indian policy or park administration anywhere in the United States. As a forcefully stated protest, however, this proposal still served as a strong counterpoint to more popular concerns about wilderness preservation and management.

In the 1970s, ideas like Iktomi Lila Sica's began to receive limited recognition from officials within the National Park Service. At Glacier, the Blackfeet redoubled their efforts to assert usufruct rights in the national park and convinced park authorities to at least waive entrance and camping fees for tribal members.[7] The Yosemite also won similar concessions from park officials at this time and played a key role in the development of a new Indian cultural museum in the valley.[8] More significantly and much closer to Iktomi Lila Sica's heart, the park service doubled the size of its holdings at Badlands National Park and expanded into the Pine Ridge Reservation. In a strong turn away from past policy, the Oglala Sioux retained ownership of all reservation land within the new park boundaries.[9] At Grand Canyon, another plan for park expansion actually led to the enlargement of the Havasupai Reservation. Furthermore, a strip of park land adjacent to the reservation became a special "traditional use area" where tribal members could, under the discretion of the secretary of the Interior, hunt and gather plants in a sustainable manner.[10]

Unlike the Yosemite Indians, the Havasupai were a federally recognized tribe with a reservation, but at least until the 1970s the experiences of native peoples in Grand Canyon and Yosemite national parks were quite similar. Though excluded from more heavily visited areas of the park—namely, Indian Gardens along the popular Bright Angel Trail and other areas on the South Rim—some Havasupai worked at Grand Canyon Village in the 1920s and 1930s. They found employment on park construction projects or worked as maids and laundresses in the hotels, and approximately ten families lived at an old village site just west of the Grand Canyon headquarters.[11] As at Yosemite, park officials considered the Indian settlement to be a dirty eyesore and relocated the Havasupai to a new village in 1934. There, the Indians paid rent to live in small cabins built by the park service, and only the gainfully employed could remain for any length of time. Weary of this small encampment, both the park service and Grand Canyon concessionaires terminated almost all of their native employees in 1955, forced them out of the village, and tore down their homes.[12]

While a couple of families remained in the park, the rest moved back to the isolated, five-hundred-acre reservation at Cataract Canyon. Located approximately thirty roadless miles from Grand Canyon Village, the area is centered on the spectacular Havasupai Falls that spill turquoise waters over red canyon walls. Though a desert oasis of unparalleled beauty, the reservation could not support a tribe of two hundred individuals and their livestock. Many Havasupai traveled off the reservation to work for wages, but others farmed in the canyon, pastured their livestock on the plateau above the reservation, and hunted or farmed surreptitiously on national park and forest lands. While forest service employees encouraged Havasupai pastoralism by reserving preferred grazing permits for Indians, the National Park Service had long kept a sharp eye on any off-reservation activities. The 1919 Grand Canyon National Park Act authorized the secretary of the Interior, "[at] his discretion, to permit individual members of [the] tribe to use and occupy other tracts of land within [the] park," but officials chose instead to restrict all native use of park lands and resources.[13]

In 1970, the park service developed a new master plan for Grand Canyon that would incorporate adjacent national forest lands into a greatly enlarged national park. Part of a new emphasis in park policy that called for the preservation of a large wilderness area, the master plan required the elimination of all use permits within the expanded boundaries. Once the Havasupai learned of these proposed changes, they appealed to powerful Arizona politicians like Senator Barry Goldwater and Representative Morris Udall. Both men worked on the tribe's behalf in Washington and sponsored legislation stipulating that any park expansion must also protect customary use of the plateau areas around Cataract Canyon. After several years of great controversy, in which groups like the Sierra Club and the National Parks Conservation Association lobbied against the "Indian threat" to Grand Canyon, the Havasupai and their allies won a decisive victory in January 1975. The reservation was increased by 185,000 acres, and the Havasupai gained "traditional usage" rights to more than 95,000 acres of adjacent national park lands. However, the tribe's victory was tempered by the success of the park service and preservationist groups in preventing the Havasupai from reestablishing their small community at Grand Canyon Village. Likewise, their rights to use national park lands still fell under the same discretionary terms of the original Grand Canyon National Park Act. Of course, the enlargement of the reservation onto the Coconino Plateau would probably never have occurred if the area possessed exceptional scenery or contained unique ecological features. Nevertheless, the Havasupai gained an exceptional victory in their ongoing conflict with the National Park Service in part because many non-Indians now found their arguments compelling.[14]

The principles that shaped events at Grand Canyon in the 1970s received further amplification when, under the outgoing administration of President Jimmy Carter, the Alaska National Interest Lands Conservation Act (ANILCA) added ten new units to the national park system and provided for subsistence use in nine of the new or enlarged land units.[15] This unique situation within the national park system grew out of a long, complicated process that predated Alaskan statehood in 1959. The act admitting Alaska to the Union allowed the state to select 103.5 million acres of federal land but also required the state to disclaim any rights or jurisdiction that might still be subject to aboriginal title. Because the United States had never signed a treaty with any Alaskan native groups and the 1884 Organic Act for the Territory of Alaska recognized aboriginal rights to lands they used or occupied, the requirement that the state respect native claims applied to virtually all land in Alaska.

Despite the filing of protective blanket claims by several native groups, the federal government began processing the state's land selections almost immediately. This mobilized and unified Alaska Natives, Eskimos, and Aleuts, who organized both regionally and statewide. Their protests eventually caused Secretary of the Interior Stewart L. Udall to halt all state land selections in 1966. Eager to exploit Alaskan mineral resources, oil company lobbyists joined in an unlikely coalition with native groups to push for speedy settlement of the state land issue. These efforts culminated in 1971 with the Alaska Native Claims Settlement Act (ANCSA), which provided for the payment of $962.5 million and conveyance of 45 million acres to Native Alaskans. Unlike aboriginal land settlements elsewhere in the United States, ANCSA did not create semisovereign reservations. A small fraction of the total native settlement did cover village sites, but the rest fell under the control of eleven native-run corporations. Although the management of valuable mineral and forest resources by these corporations seemed the best way to guarantee economic self-sufficiency, many traditionalists supported ANCSA because it also set aside large areas where they might continue older subsistence practices.[16]

When native groups were protesting the federal government's wholesale disposal of millions of acres to the state, which in turn planned to sale or lease much of the land to mining and oil development interests, national preservationist groups also lobbied for the protection of unselected lands. ANCSA reflected their interests as well, for the act stipulated that the secretary of the Interior could withdraw eighty million acres of the public domain for the creation and expansion of national parks, forests, wildlife refuges, and wild and scenic rivers. Because efforts to retain aboriginal control over certain areas presented the strongest check against massive development of mineral and energy resources, national park advocates soon joined forces with those indigenous leaders who had been pushing for the protection of traditional use areas. Their combined efforts had a strong effect on the land selection process mandated by ANCSA and soon led to the passage of ANILCA. This latter piece of legislation would link preservationist interests with native concerns in ways that profoundly shaped federal land management policies in the forty-ninth state. In Alaska, at least, it seemed possible to protect large "wilderness" areas and incorporate the cultures that had long interacted with them.[17]

Except for the example of the Havasupai and Grand Canyon National Park, the integration of subsistence use areas into some of Alaska's national parks makes them wholly unlike other protected areas in the lower forty-eight states. However, Alaska's national parks do not guarantee exclusive rights to native people. Subsistence use of national park lands applies to local communities, native or nonnative, and falls within a larger set of land use policies that permit some sport hunting and mining. The only areas within Alaska's national parks where native people have specified rights are lands owned by native corporations. But these are essentially the same rights as those held by Alaskan homesteaders with prior claims on park lands. In no instance do native rights approximate the control that Indian tribes generally exercise over reservation lands in the rest of the United States. Consequently, Alaska may be less a harbinger of future national park policies than it is a throwback to the late nineteenth century, when park officials allowed hunting and recognized the claims of homesteaders and mining interests. Where things might be headed is perhaps best indicated by the situation at Denali National Park; the enlarged and renamed successor of Mount McKinley National Park, Denali does not permit subsistence use within the original 1917 boundaries. While Denali is the only Alaskan park that approximates the symbolic importance of Glacier, Yellowstone, or Yosemite, similar restrictions also apply to Glacier Bay and Katmai—the only other park areas that were in existence before the passage of ANILCA.[18]

Perhaps the strongest challenge to the uninhabited wilderness ideal has come from the Timbisha Shoshone in Death Valley. As with so many other holdings in the national park system, the creation of Death Valley National Monument in

1933 impinged directly on the area's native inhabitants. Hunting, food gathering, and seasonal camp movements were all banned; most of the local native bands dispersed to other areas outside the monument; and the few individuals who remained were confined to a small area, where they were charged for use of the valley's scarce water resources. Of a total tribal population that now numbers approximately three hundred individuals, a few dozen families have managed to remain in a small area adjacent to the park's main tourist facilities at Furnace Creek. While most Timbisha live in communities near the national park, tribal members who reside at Furnace Creek have managed to find employment with either the park service or one of the tourist concessionaires.[19]

The Timbisha Shoshone became a federally recognized tribe in 1983, with all the privileges and sovereign rights that most tribes hold by virtue of their earlier treaties with the United States. However, they do not yet have a reservation on which to exercise their newly recognized sovereignty. The California Desert Protection Act of 1994, which enlarged and upgraded Death Valley to national park status and designated adjacent lands as federal wilderness areas, required the secretary of the Interior to conduct a study of suitable locations for a reservation. Tribal leaders have pushed for control of about a quarter of the new park and shared management of nearby wilderness areas, but park officials have vigorously resisted both efforts. Keenly aware of the precedent-setting nature of the Timbisha claims, the National Park Service fears that any major concession to the tribe would open the doors to similar claims throughout the park system.[20]

In the midst of their recent disagreements with the park service, the Timbisha Shoshone have joined with five other native groups to form the Alliance to Protect Native Rights in National Parks. The other members of the alliance are the Pai' Ohana, who have claims on Kaloko-Honokohau National Historic Park on the island of Hawaii; the Miccosukee Tribe, which has claims on Everglades National Park in Florida; the Navajo Nation, which comanages Canyon de Chelly National Monument with the park service in Arizona; the Five Sandoval Indian Pueblos, which have concerns over the management of cultural sites within Petroglyph National Monument in New Mexico; and the Hualapai Tribe, which has complained of a new "overflight" plan that affects their reservation near Grand Canyon National Park in Arizona. Although negotiations between the National Park Service and these native groups will no doubt produce a number of compromises that at least permit traditional use of some park lands, the difficult path toward resolution illustrates the persistence of century-old ideas about wilderness, land use, and native rights. Nevertheless, such agreements will probably not send any immediate tremors through the park system. Though Death Valley and these other park holdings certainly include areas of spectacular scenery and protect several endangered species, none have the same appeal as America's "crown jewel" national parks. Consequently, some degree of native use and control of these park areas would not challenge popular conceptions of wilderness in quite the same way that a group of Indian hunters in Yellowstone would.[21]

The issues surrounding native claims to Death Valley are very similar to those at Glacier National Park, where the Blackfeet first made a proposal for joint management of the eastern half of the park in 1975.[22] The Blackfeet have yet to gain recognition of the rights guaranteed them in 1895, but the park service and the tribe have begun to reach accord on how to manage certain park features that have special religious significance for Blackfeet traditionalists. For the most part, there is now goodwill on both sides, and park officials genuinely want to incorporate Blackfeet concerns into future park policy. Likewise, many Blackfeet appreciate the park service's ability to protect an area of great significance to the tribe. As one tribal elder recently observed,

> Here, our medicine on the Reservation is not as strong as that up there. And it's bigger up there. Because up there in the mountains it's so pure. And here, it's been trampled over and cars disturb it, everybody disturbing, cows graze it off. Whereas in the mountains, the elk and the wild game respect the medicines. And they use them themselves. There's so much of the Park left because it hasn't been bothered, it's been clean all these years.[23]

The park service and the Blackfeet may have a long adversarial history, but both recognize how their interests have overlapped significantly in the past few years.

Despite a shared concern for the Glacier environment, the basic issues that have divided the Blackfeet and the National Park Service for most of this century are still far from easy resolution. The impasse not only reflects the powerful cultural values that remain embedded in two very different conceptions of the same landscape but also stems from the many difficult issues that have always impinged on the exercise of tribal political sovereignty within the United States. In recent years, the park service has begun to see that native subsistence practices do not necessarily compromise the integrity of park environments, which in turn has led to greater cooperation between park superintendents and tribal councils.[24] Cultural agreements between the park service and various native groups will probably lead to the acknowledgment of past wrongs, but nothing of lasting import will take place until there is some resolution of the issue of native rights on public lands.

If cooperation on cultural issues does further tribal efforts to exercise some control over national park lands, this could revolutionize the way all Americans experience the wilderness. At Glacier National Park, for instance, full recognition of Blackfeet claims would make plain that the American preservationist ideal is predicated on Indian dispossession. Furthermore, the notion of a usable or inhabitable wilderness implies that "nature" and "culture" are deeply intertwined, if not inseparable. Rather than idolize wilderness as a nonhuman landscape, where a person can be nothing more than "a visitor who does not remain," national parks might provide important new lessons about the degree to which cultural values and actions have always shaped the "natural world."[25] More particularly, to view a national park like Glacier as part of the "head" of the Blackfeet people and not simply the "crown of the continent" might lead tourists to see themselves as visitors in Indian country and not simply as pilgrims at an American shrine. Likewise, native use of national parks like Glacier, Yellowstone, and Yosemite would further tribal efforts to reclaim their traditions and, in the process, strengthen their ability to remain politically and culturally distinct nations.

NOTES

Notes to Introduction

1. Lawrence Kip, *The Indian Council at Walla Walla, May and June, 1855* (Eugene: University of Oregon, 1897), 22.

2. John G. Neihardt, *Black Elk Speaks: Being the Life Story of a Holy Man of the Oglala Sioux* (New York: William Morrow, 1932; reprint, Lincoln: University of Nebraska Press, 1988), 9. Badlands National Monument was authorized by Congress in 1929, enlarged in 1936 and again in 1976, and redesignated Badlands National Park in 1978. For a longer but more accurate rendering of Black Elk's words, see Raymond J. DeMallie, ed., *The Sixth Grandfather: Black Elk's Teachings Given to John G. Neihardt* (Lincoln: University of Nebraska Press, 1984), 288–290.

3. Quote is from the Yellowstone Park Act, *U.S. Statutes at Large* 17(1872):32–33.

4. In some respects, Grand Canyon National Park also fits these basic criteria; see chapter 9.

5. For an excellent example of the many recent studies on how native peoples altered and "domesticated" the physical landscape, see Thomas Blackburn and Kat Anderson, eds., *Before the Wilderness: Environmental Management by Native Californians* (Menlo Park, Calif.: Ballena Press, 1993).

6. William Cronon, "The Problem with Wilderness," in *Uncommon Ground: Toward Reinventing Nature,* ed. William Cronon (New York: W. W. Norton, 1995), 78–79; William Denevan, "The Pristine Myth: The Landscape of the Americas in 1492," *Annals of the Association of American Geographers* 82 (1992): 369–385; Michael Frome, *Regreening the National*

Parks (Tucson: University of Arizona Press, 1992), 225–227; Gary Paul Nabhan, "Cultural Parallax in Viewing North American Habitats," in *Reinventing Nature? Responses to Postmodern Deconstruction,* ed. Michael E. Soulé and Gary Lease (Washington, D.C.: Island Press, 1995), 87–101; and Barbara Deutsch Lynch, "The Garden and the Sea: U.S. Latino Discourses and Mainstream Environmentalism," *Social Problems* 40 (1993): 108–125.

7. This term was used in the text for historical exhibits in Mammoth Visitor Center and Lake Yellowstone Lodge, Yellowstone National Park, summer 1997.

8. Public Law 88-577 in U.S. *Statutes at Large* 78 (1964): 890–896.

9. See, for example, Alfred Runte, *National Parks: The American Experience,* 3d ed. (Lincoln: University of Nebraska Press, 1997), 53.

10. For two excellent studies of changing land use practices on reservations, see Melissa Meyer, *The White Earth Tragedy: Ethnicity and Dispossession at a Minnesota Anishinaabe Reservation, 1889–1920* (Lincoln: University of Nebraska Press, 1995); and David Rich Lewis, *Neither Wolf Nor Dog: American Indians, Environment, and Agrarian Change* (New York: Oxford University Press, 1994).

Notes to Chapter 1

1. Henry Marie Brackenridge, *Views of Louisiana: Containing Geographical, Statistical and Historical Notices of That Vast and Important Portion of America* (Baltimore, 1817), 72; quoted in Henry Nash Smith, *Virgin Land: The American West as Symbol and Myth,* 2d ed. (Cambridge: Harvard University Press, 1970), 175.

2. George Catlin, *Letters and Notes on the Manners, Customs, and Conditions of the North American Indian* (1844); reprint, with an introduction by Marjorie Halpin, 2 vols. (New York: Dover, 1973), 1: 18; emphasis in original.

3. Ibid., 1: 3, 260.

4. Ibid., 1: 260–262; emphasis in original. This appeal for a "nation's Park" was first published in the *New York Daily Commercial Advertiser* in 1833. For an examination of Catlin's professional career, see Brian Dippie, *Catlin and His Contemporaries: The Politics of Patronage* (Lincoln: University of Nebraska Press, 1990).

5. Catlin, *Letters and Notes,* 1: 256. A number of reservations border the Missouri River in North and South Dakota, but all of these have been inundated by artificial lakes. At Fort Berthold Indian Reservation, for instance, the creation of Lake Sakakawea behind Garrison Dam in 1953 forced the relocation of 80 percent of the reservation population. See Michael L. Lawson, *Damned Indians: The Pick-Sloan Plan and the Missouri River Sioux, 1944–1980* (Norman: University of Oklahoma Press, 1982), 59–62.

6. For historical assessments of Catlin's proposal, see Hans Huth, *Nature and the American: Three Centuries of Changing Attitudes* (Berkeley: University of California Press, 1957), 134–136; Roderick Nash, *Wilderness and the American Mind,* 3d ed. (New Haven: Yale University Press, 1982), 100–107; and Alfred Runte, *National Parks: The American Experience,* 3d. ed (Lincoln: University of Nebraska Press, 1997), 26.

7. See especially John Canup, *The Emergence of an American Identity in Colonial New England* (Middletown, Conn.: Wesleyan University Press, 1990); Peter N. Carroll, *Puritanism and the Wilderness: The Intellectual Significance of the New England Frontier, 1629–1700* (New York: Columbia University Press, 1969); Annette Kolodny, *The Land Before Her: Fantasy and Experience of the American Frontiers, 1630–1860* (Chapel Hill: University of North Carolina Press, 1984); and Cecilia Tichi, *New World, New Earth: Environmental Reform in American Literature from the Puritans Through Whitman* (New Haven: Yale University Press, 1979).

8. Lee Clark Mitchell, *Witnesses to a Vanishing America: The Nineteenth Century Response* (Princeton: Princeton University Press, 1981), 93–109.

9. Brackenridge, *Views of Louisiana,* 72.

10. Nash, *Wilderness,* 47.

11. Ralph Waldo Emerson, "Nature," in *The Collected Works of Ralph Waldo Emerson: Nature, Addresses, and Lectures,* ed. Robert E. Spiller and Alfred R. Ferguson (Cambridge, Mass.: Belknap Press, 1971), 10, 37.

12. Robert F. Berkhofer Jr., *The White Man's Indian: Images of the American Indian from Columbus to the Present* (New York: Alfred A. Knopf, 1978), 71–79; Roy Harvey Pearce, *Savagism and Civilization: A Study of the Indian and the American Mind,* rev. ed. (Berkeley: University of California Press, 1988), 136–150.

13. Thomas Cole, "Essay on American Scenery," in *Thomas Cole: The Collected Essays and Prose Sketches,* ed. Marshall Tymn (St. Paul: John Colet Press, 1980), 11; first published in *American Monthly Magazine,* 1833. Cole wrote on this theme on several occasions, including the twelve-part poem "The Spirits of the Wilderness" and in his "Lecture on American Scenery: Delivered Before the Catskill Lyceum on April 1, 1841." For a more complete discussion of American romanticism, see James Thomas Flexner, *That Wilder Image: The Painting of America's Native School from Thomas Cole to Winslow Homer* (Boston: Little, Brown, 1962), 3–102; Perry Miller, "The Romantic Dilemma in American Nationalism and the Concept of Nature," *Harvard Theological Review* 48 (1955): 239–253; Barbara Novak, *Nature and Culture: American Landscape and Painting, 1827–1875,* rev. ed. (New York: Oxford University Press, 1995), 3–45; and Max Oelschlager, *The Idea of Wilderness: From Prehistory to the Age of Ecology,* (New Haven: Yale University Press, 1991), 68–133.

14. Emerson, "The American Scholar," in *Collected Works,* 69.

15. Quotations are from Lydia Sigourney's "Meeting of the Susquehanna with the Lackawanna" and Lucretia Davidson's "America." Sigourney published more than sixty books of prose and verse; Davidson became something of a cult figure after her death at the age of seventeen. See Joanne Dobson, *Dickinson and the Strategies of Reticence: The Woman Writer in Nineteenth-Century America* (Bloomington: Indiana Universtiy Press, 1989), 76–85; and Lucretia Maria Davidson, *Poetical Remains of the Late Lucretia Maria Davidson, Collected and Arranged by Her Mother; with a Biography by Miss Sedgewick,* (Philadelphia: Lea and Blanchard, 1846).

16. Smith, *Virgin Land,* 49–89.

17. Tichi, *New World, New Earth,* ix; Rayna Diane Green, "The Only Good Indian: The Image of the Indian in American Vernacular Culture" (Ph.D. diss., Indiana University, 1973).

18. James D. Wallace, *Early Cooper and His Audience* (New York: Columbia University Press, 1986), 171.

19. For a more complete discussion of Cooper's works, see Geoffrey Rans, *Cooper's Leather-Stocking Novels: A Secular Reading* (Chapel Hill: University of North Carolina Press, 1991); and Robert Clark, ed., *James Fenimore Cooper: New Critical Essays* (London: Vision Press, 1985).

20. Christine Stansell and Sean Wilentz, "Cole's America," in *Thomas Cole: Landscape into History,* ed. William Truettner and Alan Wallach (New Haven: Yale University Press, 1994).

21. Alexis de Tocqueville, *Democracy in America,* trans. Henry Reeve, revised by Frances Bowen and Phillip Bradley, 2 vols. (New York: Alfred A. Knopf, 1963), 1: 118.

22. Cole, "Essay on American Scenery," 17. For a broader discussion of these sentiments, see Mitchell, *Witnesses,* 2–21.

23. Ronald N. Satz, *American Indian Policy in the Jacksonian Era* (Lincoln: University of Nebraska Press, 1975); Michael Paul Rogin, *Fathers and Children: Andrew Jackson and the Subjugation of the American Indian* (New York: Alfred A. Knopf, 1975); Francis Paul Prucha, *Indian Policy in the United States: Historical Essays* (Lincoln: University of Nebraska Press, 1981), 92–116, 138–142; Lynn Hudson Pasons, "'A Perpetual Harrow on My Feelings': John Quincy Adams and the American Indian," *New England Quarterly* 46 (1973): 339–379.

24. Black Hawk, *Black Hawk: An Autobiography,* ed. Donald Jackson (Chicago: University of Illinois Press, 1955); Virginia Bergman Peters, *The Florida Wars* (Hamden, Conn.: Archon Books, 1975); Anthony F. C. Wallace, *Prelude to Disaster: The Course of Indian-White Relations Which Led to the Black Hawk War of 1832* (Springfield: Illinois State Historical Library, 1970); and Thurman Wilkins, *Cherokee Tragedy: The Ridge Family and the Decimation of a People,* 2d rev. ed. (Norman: University of Oklahoma Press, 1986). The most thorough examination of the various legal issues surrounding Cherokee sovereignty is Jill Norgren's *The Cherokee Cases: The Confrontation of Law and Politics* (New York: McGraw Hill, 1996).

25. Quotation is from Cooper, *Notions of the Americans: Picked Up by a Travelling Bachelor* (Albany: State University of New York Press, 1991), 682; originally published in 1828.

26. Satz, *American Indian Policy,* 211–236.

27. Preston Holder, *The Hoe and the Horse on the Plains: A Study of Cultural Development among North American Indians* (Lincoln: University of Nebraska Press, 1970); James Clifton, *The Prairie People: Continuity and Change in Potawatomi Indian Culture, 1665–1965* (Lawrence: Regents Press of Kansas, 1977), 279–342; R. David Edmunds, *The Potawatomis: Keepers of the Fire* (Norman: University of Oklahoma Press, 1978), 240–272; Richard White, "The Winning of the West: The Expansion of the Western Sioux in the Eighteenth and Nineteenth Centuries," *Journal of American History* 65 (1978): 319–343; Bertha Waseskuk, "Mesquakie History–As We Know It," in *The Worlds Between Two Rivers: Perspectives on American Indians in Iowa,* ed. Gretchen M. Bataile et al. (Ames: Iowa State University Press), 54–61.

28. A. R. Fulton, *The Red Men of Iowa* (Des Moines: Mills, 1882), 238.

29. Francis La Flesche, *The Middle Five: Indian Schoolboys of the Omaha Tribe* (Madison: University of Wisconsin Press, 1963), xx.

30. Quotations are from the Indian Removal Act of 1830; reprinted in Francis Paul Prucha, ed., *Documents of United States Indian Policy* (Lincoln: University of Nebraska Press, 1975), 52–53.

31. Earl Arthur Shoemaker, *The Permanent Indian Frontier: The Reason for the Construction and Abandonment of Fort Scott, Kansas, During the Dragoon Era* (n.p.: National Park Service, 1986).

32. Washington Irving, *The Western Journals of Washington Irving,* ed. John Francis McDermott (Norman: University of Oklahoma Press, 1944), 10, 12. Irving's experiences were later published in *A Tour of the Prairies* (1832), which was essentially a polished version of his detailed journals and letters.

33. Washington Irving, *The Adventures of Captain Bonneville* (Boston: Twayne, 1977), 269. Also see Nash, *Wilderness,* 98.

34. Maria Rebecca Audubon, *Audubon and His Journals,* 2 vols. (New York: Charles Scribner and Sons, 1897), 1: 374, 378–79, 406–407.

35. John James Audubon, *Delineations of American Scenery and Character* (New York: G. A. Baker, 1926), 4, 9.

36. Audubon, *Audubon and His Journals,* 2: 131; 1: 488, 496–497. The great auk, which was extinct by 1844, had been decimated by commercial egg collectors along the islands of the North Atlantic coast of North America. Audubon had personally witnessed this destruction in 1832 and wrote a number of vitriolic essays damning the "eggers of Labrador."

37. Osborne Russell, *Osborne Russell's Journal of a Trapper,* ed. Aubrey L. Haines (Lincoln: University of Nebraska Press, 1965), 26–27, 46. Russell's "Secluded Valley" is now the Lamar Valley in the northeastern section of Yellowstone National Park. For a sample of the many popular books written by fur trappers, see Reuben Ford Thwaites, ed., *Early Western Travels: 1748–1846,* 32 vols. (Cleveland: Arthur H. Clark, 1904–1907); and Henry R. Wagner, *The Plains and the Rockies: A Bibliography of Original Narratives of Travel and Adventure, 1800–1865* (San Francisco: Grabhorn Press, 1937). Also see W. A. Ferris, *Life in the Rocky Mountains: A*

Diary of Wanderings on the Sources of the Rivers Missouri, Columbia, and Colorado from February, 1830, to November 1835, ed. Paul C. Phillips (Denver: Old West Publishing, 1940).

38. Petition of the Amoskeag workers quoted in Herbert G. Gutman, *Work, Culture, and Society in Industrializing America* (New York: Alfred A. Knopf, 1976), 29.

39. Charles Lanman, *A Summer in the Wilderness; Embracing a Canoe Voyage Up the Mississippi and Around Lake Superior* (New York: D. Appleton, 1847), 128, emphasis added; also see 91, 100–103, 120–123. For a brief biography of Lanman, see Nash, *Wilderness,* 62, 97.

40. Charles Lanman, *The Adirondacks* (New York: D. Appleton, 1854), 48, 107.

41. The foregoing discussion of Thoreau is greatly influenced by the following works: William Howarth, *Thoreau in the Mountains* (New York: Farrar, Straus, Giroux, 1982); Max Oelschlager, *The Idea of Wilderness,* 133–171; Robert F. Sayre, *Thoreau and the American Indians* (Princeton: Princeton University Press, 1977); and Elizabeth I. Hanson, *Thoreau's Indian of the Mind* (Lewiston, N.Y.: Edwin Mellen Press, 1991).

42. Henry David Thoreau, *The Maine Woods* (New York: Harper and Row, 1987), 92.

43. Ibid., 93, 95, emphasis in original; Emerson, "Nature," 21.

44. Thoreau, *Maine Woods,* 97. For an argument that Thoreau's experience in Maine caused him to be more critical in his ideas about Native Americans, see Nash, *Wilderness,* 46.

45. Thoreau, *Walden and Other Writings by Henry David Thoreau,* ed. Joseph Wood Krutch (New York: Bantam, 1962), 172, 187.

46. Thoreau quoted in Richard F. Fleck, *Henry Thoreau and John Muir Among the Indians* (Hamden, Conn.: Archon Books, 1985), 17.

47. Thoreau, "Walking," in *The Writings of Henry David Thoreau,* 20 vols. (Boston: Houghton Mifflin, 1906), 5: 218. This quote has become the motto of the Wilderness Society.

48. Thoreau, "Chesuncook," *Atlantic Monthly* 2 (1858): 317; quoted in Huth, *Nature and the American,* 169.

49. For an examination of Thoreau's lifelong interest in Indians, see Sayre, *Thoreau and the American Indians.*

50. John Muir, *My First Summer in the Sierra* (Boston: Houghton Mifflin, 1916), 226.

Notes to Chapter 2

1. Luther Standing Bear, *Land of the Spotted Eagle* (Boston: Houghton Mifflin, 1933), xix.

2. Samuel Bowles, *Across the Continent: A Summer's Journey to the Rocky Mountains, the Mormons, and the Pacific States, with Speaker Colfax* (Springfield, Mass.: Samuel Bowles, 1865), ii–iv; and Anne Farrar Hyde, *An American Vision: Far Western Landscape and National Culture, 1820–1920* (New York: New York University Press, 1990), 70.

3. Bowles, *The Parks and Mountains of Colorado: A Summer Vacation in the Switzerland of America, 1868,* ed. James H. Pickering (Norman: University of Oklahoma Press, 1991), 35–36; originally published as *The Switzerland of America* (1869). What Bowles called the Hot Springs Valley is now the small town of Hot Sulphur Springs, a few miles southeast of Rocky Mountain National Park.

4. Ibid., 94. At the time of his visit, the upper branch of the Colorado River was then called the Grand River

5. See chapter 1.

6. Bowles, *Parks and Mountains of Colorado,* 97, 103, 182.

7. First established in 1864 by congressional act, the lands comprising Yosemite Park were removed from the public domain and turned over to the state of California. Yosemite National Park was not established until 1890 and did not include the state-administered lands until 1906. See chapter 7.

8. Bowles, *Across the Continent,* 231. Bowles, *Our New West* (Hartford, Conn.: Hartford Publishing, 1869), 385. Also see Alfred Runte, *National Parks: The American Experience,* 3d ed. (Lincoln: University of Nebraska Press, 1997), 11–13.

9. Bowles, *Parks and Mountains of Colorado,* 182. Rocky Mountain National Park was not established until 1915, but Bowles should at least be credited as one of the first advocates for a national park in the Colorado Rockies. Though neither discusses him at any length, the two standard histories on the park are C. W. Buchholtz, *Rocky Mountain National Park: A History* (Boulder: Colorado Associated University Press, 1983); and Karl Hess, *Rocky Times in Rocky Mountain National Park: An Unnatural History* (Niwot: University Press of Colorado, 1993).

10. Bowles, *Parks and Mountains of Colorado,* 145–147. While he did not have as fatalistic a view of native peoples' futures, Commissioner Ely S. Parker, who was himself a Seneca chief, long advocated the cessation of treaties for many of the same reasons that Bowles lists; see *Annual Report of the Commissioner of Indian Affairs Made to the Secretary of the Interior for the Year 1869* (Washington, D.C.: Government Printing Office, 1870), 6 (hereafter *ARCIA*). The policy of entering into treaties with Indian tribes was abolished by Congress in 1871.

11. Bowles, *Parks and Mountains of Colorado,* 147–148.

12. Though his views seemed increasingly eccentric, it should be noted that the aged George Catlin continued to restate many of his old arguments through the late 1860s. See George Catlin, *Last Rambles Amongst the Indians of the Rocky Mountains and the Andes* (London: Sampson Low & Son, and Marston, 1868). For a discussion of Catlin's last years, see Brian W. Dippie, *Catlin and His Contemporaries: The Politics of Patronage* (Lincoln: University of Nebraska Press, 1990), 359–382.

13. John L. O'Sullivan, quoted in Julius W. Pratt, "The Origins of Manifest Destiny," *American Historical Review* 32 (1927): 796.

14. Reginald Horsman, *Race and Manifest Destiny: The Origins of American Racial Anglo-Saxonism* (Cambridge: Harvard University Press, 1981), 229–248.

15. For the "discovery" of Yosemite in 1851 and its immediate consideration as a "natural monument" for the entire nation, see Lafayette H. Bunnell, *Discovery of the Yosemite, and the Indian War of 1851, Which Led to That Event* (Chicago: Fleming H. Revel, 1880).

16. Horace Greeley exactly expressed these later sentiments in *An Overland Journey from New York to San Francisco in the Summer of 1859* (New York: C. M. Saxton, Barker, 1860), 311–312.

17. Robert Utley, *The Indian Frontier of the American West, 1846–1890* (Albuquerque: University of New Mexico Press, 1984), 1–47; Martha Royce Blaine, *The Ioway Indians* (Norman: University of Oklahoma Press, 1979), 205–266; Joseph B. Herring, *The Enduring Indians of Kansas: A Century and a Half of Acculturation* (Lawrence: University Press of Kansas, 1990), 119–146; and H. Craig Miner and William E. Unrau, *The End of Indian Kansas* (Lawrence: University Press of Kansas, 1978).

18. Utley, *Indian Frontier,* 52-81. By "popular," I mean the unofficial maps in overland guidebooks and travelogues. For a general discussion of nineteenth-century maps of the West, see Walter W. Ristow, *American Maps and Mapmakers: Commercial Cartography in the Nineteenth Century* (Detroit: Wayne State University Press, 1985), 445–466.

19. Melville's phrase is the title for a chapter of the novel. See Roy Harvey Pearce, *Savagism and Civilization: A Study of the Indian and the American Mind,* rev. ed. of *The Savages of America,* 1953 (Berkeley: University of California Press, 1988), 244–251; and Richard Drinnon, *Facing West: The Metaphysics of Indian Hating and Empire Building,* 2d ed. (New York: Schocken, 1990), 191–215. For a broader analysis of white attitudes toward Indians in the midnineteenth century, see Robert Berkhofer Jr., *The White Man's Indian: Images of the Ameri-*

can Indian from Columbus to the Present (New York: Vintage Books, 1979); and Rayna Diane Green, "The Only Good Indian: The Image of the Indian in American Vernacular Culture" (Ph.D. diss., Indiana University, 1973).

20. Quote is from the *Annual Report of the Commissioner of Indian Affairs for the Year 1851,* 32d Cong., 1st sess., 1851, H. Doc. 2, serial 636, p. 2. The standard work on the early development of the reservation system is still Robert A. Trennert Jr., *Alternative to Extinction: Federal Indian Policy and the Beginnings of the Reservation System* (Philadelphia: Temple University Press, 1975). Also see Edmund Jefferson Danziger Jr., *Indians and Bureaucrats: Administering the Reservation Policy During the Civil War* (Urbana: University of Illinois Press, 1974); Utley, *Indian Frontier,* 31–63; and Francis Paul Prucha, *The Great Father: The United States Government and the American Indians,* 2 vols. (Lincoln: University of Nebraska Press, 1984), 1: 192–211.

21. Richard White, *"It's Your Misfortune and None of My Own": A History of the American West* (Norman: University of Oklahoma Press, 1991), 92.

22. *ARCIA, 1858,* 6–7. For analysis of the failed reservation system in California, see George Harwood Philips, *Indians and Indian Agents: The Origins of the Reservation System in California, 1849–1852* (Norman: University of Oklahoma Press, 1997); and Albert L. Hurtado, *Indian Survival on the California Frontier* (New Haven: Yale University Press, 1988), 100–124. For conflicts and policy failures in both the Pacific Northwest and Texas, see Utley, *Indian Frontier,* 52–56.

23. These treaty councils took place in 1851 at Fort Laramie, Wyoming Territory; in 1853, at Fort Atkinson, Kansas; and in 1855 at the mouth of the Judith River, Montana Territory. For the text of these ratified agreements, see Charles J. Kappler, comp., *Indian Affairs: Laws and Treaties,* 5 vols. (Washington, D.C.: Government Printing Office, 1904–1941) 2: 594–596, 600–602, and 736–740. Also see Raymond J. DeMallie, "Touching the Pen: Plains Indian Treaty Councils in Ethnohistorical Perspective," in *Ethnicity on the Great Plains,* ed. Frederick C. Luebke (Lincoln: University of Nebraska Press, 1980), 38–51.

24. For an excellent overview of the environmental destruction wrought by emigrant trains, see Elliott West, *The Way to the West: Essays on the Great Plains* (Albuquerque: University of New Mexico Press, 1995), 27–38. Also see Utley, *Indian Frontier,* 65–127. Quote is from the Fort Laramie Treaty of 1851 in Kappler, *Indian Affairs,* 2: 594.

25. Susan Badger Doyle, "Indian Perspectives of the Bozeman Trail, 1864–8," *Montana, The Magazine of Western History* 40 (1990): 56–67; Thomas Dunlay, *Wolves for the Blue Soldiers: Indian Scouts and Auxiliaries with the United States Army, 1860–1890* (Lincoln: University of Nebraska Press, 1982). For Sioux power on the plains, see Richard White, "The Winning of the West: The Expansion of the Western Sioux in the Eighteenth and Nineteenth Centuries," *Journal of American History* 65 (1978): 319–343.

26. Blackfoot, also known as Sits in the Middle of the Land, is quoted in Louis L. Simonin, *The Rocky Mountain West in 1867,* trans. Wilson O. Clough (Lincoln: University of Nebraska Press, 1966), 110–111; also see *Papers Relating to Talks and Councils Held with the Indians in Dakota and Montana Territories in the Years 1866–1869* (Washington, D.C.: Government Printing Office, 1910), 68–69. The 1867 meeting was preliminary to official treaty negotiations conducted at Fort Laramie in May 1868.

27. Quotation of the 1867 law creating the Taylor Peace Commission reprinted in Francis Paul Prucha, ed., *Documents of United States Indian Policy* (Lincoln: University of Nebraska Press, 1975), 106.

28. Quote is part of article 4 in the 1868 Fort Laramie Treaty with the Crow; see Kappler, *Indian Affairs,* 2: 1009. The exact same phrase is used in article 4 of a treaty negotiated with the Shoshone and Bannock at Fort Bridger, Idaho, in July 1868; see Kappler, *Indian Affairs,* 2: 1021. Similar clauses were included in treaty negotiations with the Sioux and Arapaho in April and the Northern Cheyenne and Northern Arapaho in May; see Kappler,

Indian Affairs, 2: 1002, 1012. Also see *Papers Relating to Talks and Councils,* 95–108. For excerpts from the proceedings of the 1867 Medicine Lodge Creek Treaty councils with the Kiowa, Comanche, Cheyenne, and Arapaho, see Douglas C. Jones, *The Treaty of Medicine Lodge: The Story of the Great Treaty Council as Told by Eyewitnesses* (Norman: University of Oklahoma Press, 1966). These various negotiations are part of the last group of treaties between the United States and Indian tribes.

29. Lieutenant General Sherman to E[dwin]. M. Stanton, May 8, 1868; in 40th Cong., 2d sess., 1868, H. Doc. 239, serial 1341, p. 2.

30. Kappler, *Indian Affairs,* 2: 1008.

31. Washakie quoted in Virginia Cole Trenholm and Maurine Carley, *The Shoshonis: Sentinels of the Rockies* (Norman: University of Oklahoma Press, 1964), 222. Similar statements can be found in *Papers Relating to Talks and Councils,* 56–72, 95–108. For an examination of the different ways that native leaders and government commissioners understood treaty negotiations, see DeMallie, "Touching the Pen."

32. Utley, *Indian Frontier,* 173–189.

33. Horace Greeley quoted in Roderick Nash, *Wilderness and the American Mind,* 3d ed. (New Haven: Yale University Press, 1982), 96.

34. Bayard Taylor, *At Home and Abroad: A Sketch Book of Life, Scenery, and Men,* 2d ed. (New York: G. P. Putnam, 1862), 155.

35. Runte, *National Parks,* 29–32.

36. Charles Joseph Latrobe, The *Rambler in North America: 1832–1833* (London: R. B. Seeley and W. Burnside, 1835), 73–74.

37. For a broader discussion of the importance of Niagara Falls to a national self-image, see Elizabeth McKinsey, *Niagara Falls: Icon of the American Sublime* (New York: Cambridge University Press, 1985). Runte also comments on the importance of Niagara for the creation of the first national parks, *National Parks,* 5–9.

38. Bowles, *Parks and Mountains of Colorado,* 181–182.

39. William H. H. Murray, *Adventures in the Wilderness; or Camplife in the Adirondacks* (Boston: Fields, Osgood, 1869), 8. For a general discussion of the growing popularity of outdoor recreation in the 1860s, see Hans Huth, *Nature and the American: Three Centuries of Changing Attitudes* (Berkeley: University of California Press, 1957), 96–102. Huth does not discuss the importance of "camp-life" for women at this time, but Murray comments extensively on outdoor recreation for "the ladies." A large part of women's fascination with the outdoors and the social acceptability of women hikers and campers stemmed from the tremendous vogue of natural history among women and young couples in the midnineteenth century; see Lynn Barber, *The Heyday of Natural History, 1820–1870* (Garden City, N.Y.: Doubleday, 1980). For an excellent analysis of the class-based concerns of these early outdoor enthusiasts, see Karl Jacoby, "Class and Environmental History: Lessons from 'The War in the Adirondacks,'" *Environmental History* (1997): 324–342.

40. Bowles, *The Parks and Mountains of Colorado,* 181, 183. Like Murray in the Adirondacks, Bowles and several other travelers met almost an equal number of women and men in Middle Park. See also 95–96, 182; George W. Pine, *Beyond the West,* 2d ed. (Utica, N.Y.: T. J. Griffiths, 1871), 259; and Wallace Stegner, *Beyond the Hundredth Meridian: John Wesley Powell and the Second Opening of the West* (Boston: Houghton Mifflin, 1954), 30–34.

41. George Perkins Marsh, *Man and Nature; or Physical Geography as Modified by Human Action,* ed. David Lowenthal (Cambridge: Harvard University Press, Belknap Press, 1965), 43. For an analysis of Marsh's life and the impact of his work, see David Lowenthal, *George Perkins Marsh: Versatile Vermonter* (New York: Columbia University Press, 1958).

42. Marsh, *Man and Nature,* 29, 35, 36, 203–204.

43. For discussions of Marsh's relevance to later environmental thinkers, see Huth,

Nature and the American, 171–177; Lee Clark Mitchell, *Witnesses to a Vanishing America: The Nineteenth Century Response* (Princeton: Princeton University Press, 1981), 61–62; Nash, *Wilderness,* 104–105. The most detailed analysis of Marsh's influence on his contemporaries is Lowenthal, *George Perkins Marsh,* 246–268.

44. For Billings's ideas about Yosemite, Yellowstone, his indebtedness to Marsh, and his career with Northern Pacific, see Robin W. Winks, *Frederick Billings: A Life* (New York: Oxford University Press, 1991).

45. Ibid., 277.

46. Ibid. 291. Lignite coal, which was cheap and plentiful along the Northern Pacific line, caused more fires than other fuels because it required one man to shovel coal and another to simultaneously remove hot ashes, which often blew onto the prairies and plains. For Billings's ideas about Indians, see ibid., 231–236.

47. George P. Belden, *Belden, The White Chief; or Twelve Years Among the Wild Indians of the Plains* (Cincinnati: C. F. Vent, 1870), 20.

48. Ibid., 92, 166.

49. Ibid., 438.

50. Ibid., 436–448. Against protracted resistance from Crow leaders, Big Horn Canyon was flooded behind Yellowtail Dam in the 1950s. The area has since been designated a national recreation area.

51. From 1850 to 1900, Indian numbers declined by 37 percent, from approximately 400,000 to about 250,000. By contrast, the United States disposed of more than 50 percent of the public domain over the same period. See Russell Thornton, *American Indian Holocaust and Survival: A Population History since 1492* (Norman: University of Oklahoma Press, 1987), 91–133; and Roy Robbins, *Our Landed Heritage: The Public Domain, 1776–1936* (Lincoln: University of Nebraska Press, 1962), 182.

52. Herman Haupt, *The Yellowstone National Park* (New York: J. M. Stoddart, 1883), 5.

53. [Ferdinand V. Hayden], "The Wonders of the West II: More about Yellowstone," *Scribner's Monthly,* February 1872, 396.

54. *U.S. Statutes at Large* 392, (1872) 32–33.

55. For the belief that Indians could not live in Yellowstone, see the statement of Henry L. Dawes in *Congressional Globe,* 42d Cong., 2d sess., 1872, 1243; quoted in Runte, *National Parks,* 53. Dawes was the sponsor of the bill and a personal friend of both Samuel Bowles and Frederick Billings. He would gain his greatest fame as a strong advocate for the aggressive assimilation of Indians into American society.

Notes to Chapter 3

1. "Report of the Commission to Negotiate with the Crow Tribe of Indians," in *Annual Report of the Commissioner of Indian Affairs to the Secretary of the Interior, 1873* (Washington, D.C.: Government Printing Office, 1873), 127 (hereafter *ARCIA*).

2. Nathaniel Pitt Langford, *The Discovery of Yellowstone National Park: Journal of the Washburn Expedition to the Yellowstone and Firehole Rivers in the Year 1870,* with a forward by Aubrey L. Haines (Lincoln: University of Nebraska Press, 1972), 117–118. The famous campfire discussion of Yellowstone's future is almost certainly a myth; Paul Schullery and Lee Whittlesey, "The Madison Campfire Story: Yellowstone's Creation Myth and Its Legacy" (paper presented at People and Place: The Human Experience in Greater Yellowstone; fourth biennial conference on the Greater Yellowstone ecosystem, Mammoth, Wyo., Yellowstone National Park, October 1997), copy in possession of the author (hereafter, People and Place Conference).

3. See, for example, Merrill D. Beal, *The Story of Man in Yellowstone* (Caldwell, Idaho:

Caxton Printers, 1949), 131–134; Richard Bartlett, *Nature's Yellowstone* (Albuquerque: University of New Mexico Press, 1974), 166–181; and Roderick Nash, *Wilderness and the American Mind,* 3d ed. (New Haven: Yale University Press, 1982), 109–110.

4. Though one member was lost and the group occasionally divided camps, the entire Washburn expedition consisted of nine "gentlemen of standing," a six-member military escort, two cooks, and two packers. The largest previous group consisted of fourteen prospectors who entered the area for a short while in the summer of 1864. The previous year a group of forty prospectors approached the Yellowstone area, but after splintering into several smaller contingents only a few of their number actually spent time within the present boundaries of the park. For accounts of early visits to Yellowstone, see Aubrey L. Haines, *Yellowstone National Park: Its Exploration and Establishment* (Washington, D.C.: National Park Service, 1974), 1–47; and Lee H. Whittlesey, "Visitors to Yellowstone Hot Springs Before 1870," *The Journal of the Geyser Observation and Study Association* 4 (1993): 203–211.

5. Aside from their own high-powered rifles, the expedition's six-man military attachment carried one hundred rounds of ammunition each; Gustavus C. Doane, "The Report of Gustavus C. Doane upon the So-Called Yellowstone Expedition of 1870 to the Secretary of War," reprinted in Louis C. Cramton, *Early History of Yellowstone National Park and Its Relation to National Park Policies* (Washington, D.C.: Government Printing Office, 1932), 113.

6. Langford, *Discovery of Yellowstone,* 70.

7. Cornelius Hedges, "Mount Everts," Helena Daily Herald, October 8, 1870; reprinted in Cramton, *Early History,* 97; Langford, *Discovery of Yellowstone,* 153; and Doane, "Yellowstone Expedition of 1870," 137.

8. Walter Trumbull, "The Washburn Yellowstone Expedition," *Overland Monthly* 6 (1871): 436. Also see Langford, *Discovery of Yellowstone,* 97; and Doane, "Yellowstone Expedition of 1870."

9. Doane, "Yellowstone Expedition of 1870," 130. Doane used these words to describe an island in Yellowstone Lake. The following summer, another exploring party would visit the island, where they found numerous game tracks and signs of native habitation; Ferdinand V. Hayden, *Preliminary Report of the U.S. Geological Survey of Montana and Portions of Adjacent Territories; Being a Fifth Annual Report of Progress* (Washington, D.C.: Government Printing Office, 1872), 95–97.

10. Harold McCracken, comp., *The Mummy Cave Project in Northwestern Wyoming* (Cody, Wyo.: Buffalo Bill Historical Center, 1978); K. P. Cannon and K. L. Pierce, "Caldera Unrest and Human Settlement: The Prehistory of Yellowstone Lake," paper presented at People and Place Conference. For brief overviews of Yellowstone's ancient human history, see Åke Hultkrantz, "The Indians in Yellowstone Park," *Annals of Wyoming* 29 (1957): 129–133; Alston Chase, *Playing God in Yellowstone: The Destruction of America's First National Park* (Boston: Atlantic Monthly Press, 1986), 95–102; B. Robert Butler, *A Guide to Understanding Idaho Archaeology* (Pocatello: Idaho State University Museum, 1968), 32–43; and Joel C. Janetski, *Indians of Yellowstone Park* (Salt Lake City: University of Utah Press, 1987), 13–25.

11. More commonly known as Mound Builders, Hopewellian people lived in the Ohio River valley between 200 and 500 A.C.E.; for a study of Yellowstone obsidian at Hopewellian archaeological sites, see James B. Griffen, A. A. Gordus, and G. A. Wright, "Identification of Hopewellian Obsidian in the Middle West," *American Antiquity* 34 (1969): 1–14. Like their Hopewellian predecessors, Mississippian peoples depended on wide-ranging trade networks and built large mound cities such as Cahokia, near present-day St. Louis. Reaching its zenith by the eleventh century, Mississippian culture had almost completely disappeared by the fifteenth century; for evidence of Mississippian materials found in Yellowstone, see Dee C. Taylor, "Preliminary Archaeological Investigations in Yellow-

stone National Park" (1964), 70–73, manuscript on file in Yellowstone National Park Archives, Mammoth, Wyo., Yellowstone National Park (hereafter YNPA).

12. William B. Sanborn, "Indian Artifacts at the Norris Geyser Basin," and Samuel M. Beal, "Indian Camps in the Lower Geyser Basin," *Yellowstone Nature Notes* 23 (1949): 6–8 and 10–11.

13. Magdalen Moccasin, interview with the author, Crow Agency, Mont., July 7, 1994; Hayman Wise, interview with Joseph Weixelman, Old Faithful, Yellowstone National Park, October 17, 1991, tapes and transcripts in YNPA.

14. For mention of the arrowhead found in the hot spring, see Orin H. Bonney and Lorraine Bonney, *Battle Drums and Geysers: The Life and Journals of Lt. Gustavus Cheyney Doane, Soldier and Explorer of the Yellowstone and Snake River Regions* (Chicago: Swallow Press, 1970), 172. For a version of the Salish story, see Ella E. Clark, *Indian Legends from the Northern Rockies* (Norman: University of Oklahoma Press, 1966), 86–90. Åke Hultkrantz speculates on the spiritual powers that native people would have attributed to Yellowstone's geysers, but his conclusion that Indians must have feared the area contradicts historical and archaeological evidence; see Hultkrantz, "The Indians and the Wonders of Yellowstone," *Ethnos* 19 (1954): 34–68. The best study of native use and perceptions of Yellowstone's geysers is Joseph Weixelman, "The Power to Evoke Wonder: Native Americans and the Geysers of Yellowstone National Park" (master's thesis: Montana State University, 1992). For evidence of Crow use of the Yellowstone hot springs to boil meat, see Edwin Thompson Denig, *Five Indian Tribes of the Upper Missouri: Sioux, Arickaras, Assiniboines, Crees, Crows,* ed. John C. Ewers (Norman: University of Oklahoma Press, 1961), 141.

15. William Henry Jackson, *Photographs of the Yellowstone National Park and Views in Montana and Wyoming Territories* (Washington, D.C.: Government Printing Office, 1873); text accompanies plate 5.

16. Francis Stewart, interview with the author, Wyola, Mont., July 9, 1994; Kerry Stewart, trans. For a discussion of Shoshone vision quests in Yellowstone, see Joseph Weixelman, "The Power to Evoke Wonder," 54–57.

17. For a general study of aboriginal fire use in North America, see Stephen J. Pyne, *Fire in America: A Cultural History of Wildland and Rural Fire* (Princeton: Princeton University Press, 1982); a more specific study of native burning in alpine environments is Nancy Turner's "Burning Mountainsides for Better Crops: Aboriginal Landscape Burning in British Columbia," *Archaeology in Montana* 32 (1991): 57–73. For the impact of native-caused fires in Yellowstone, see Charles E. Kay, "Aboriginal Overkill and Native Burning: Implications for Modern Ecosystem Management," in *Sustainable Society and Protected Areas: Contributed Papers of the 8th Conference on Research and Resource Management in Parks and on Public Lands,* ed. Robert M. Linn (Hancock, Mich.: George Wright Society, 1995). For comments on Yellowstone's "open district[s]" in 1871 and the difficulty of traveling outside burned-over areas, see Doane, "Yellowstone Expedition of 1870," 131, 144–145.

18. Doane, "Yellowstone Expedition of 1870," 116–117; Langford, *Discovery of Yellowstone,* 72. Philetus W. Norris may have seen similar fires in the northeastern part of the future park in June 1870: "Journals," letter no. 2, March 15, 1876, Philetus Norris Collection, Henry E. Huntington Library, San Marino, Calif.

19. At present, there exist no studies of human-dependent plants that may have disappeared from Yellowstone. For studies of the near extinction of native plant foods in other protected alpine areas, see Kat Anderson, "Native Californians as Ancient and Contemporary Cultivators" in *Before the Wilderness: Environmental Management by Native Californians,* ed. Thomas C. Blackburn and Kat Anderson (Menlo Park, Calif.: Ballena Press, 1993), 167–174.

20. Despite the relatively amorphous nature of individual and band identity, there remains a broad but firm distinction between the Shoshone of present-day Idaho and

Wyoming and the Western or Basin-Plateau Shoshone of Utah, Nevada, and southeastern California. Originally distinguished by regional traits, many of the Eastern and Northern Shoshone acquired horses in the eighteenth century and became even less affiliated with their nonequestrian relatives to the west. For an overview of community identities within larger Shoshone culture, see Robert F. Murphy and Yolanda Murphy, "Shoshone-Bannock Subsistence and Society," in *Anthropological Records,* 16:7 (Berkeley: University of California, 1960), 293–338.

21. The mummified remains of a man dressed in a manner similar to mountain-dwelling Shoshone suggest they may have inhabited the area for at least the past two thousand years; see McCracken, *The Mummy Cave Project.* Also Larry Loendorf, "Prehistory of the Sheep Eater Indians," paper presented at People and Place Conference.

22. Demetri Boris Shimkin, "Shoshone-Comanche Origins and Migrations," in *Proceedings of the Sixth Pacific Science Congress* (Berkeley: University of California Press, 1939), 17–25.

23. The Lewis and Clark expedition first encountered the Shoshone near the Continental Divide in 1805, about 120 miles west of the present park boundaries. Led by Sacagawea's brother Cameahwait, this particular band of some two or three hundred people had recently suffered at the hands of the Blackfeet. See James P. Ronda, *Lewis and Clark among the Indians* (Lincoln: University of Nebraska Press, 1984), 143–154.

24. Though never an important locale for buffalo hunters, the park area often harbored sizable herds that no doubt proved important to both equestrian and nonequestrian hunters. In June 1870, a party of prospectors encountered a herd of several thousand buffalo in the northeastern portion of the park; see Haines, *Yellowstone National Park,* 39–40.

25. For a discussion of Shoshone participation in these annual rendezvous, see Virginia Cole Trenholm and Maurine Carley, *The Shoshonis: Sentinels of the Rockies* (Norman: University of Oklahoma Press, 1964), 56–73.

26. Demetri Boris Shimkin, "Wind River Shoshone Ethnogeography," *Anthropoligical Records,* vol. 5, no. 4 (Berkley: University of California, 1947) 279; Murphy and Murphy, "Shoshone-Bannock Subsistence and Society," 305–310.

27. While all writers on the Northern and Eastern Shoshone have discussed the Bannock, two studies that focus exclusively on the latter group are Omer C. Stewart, "The Question of Bannock Territory," in *Languages and Cultures of Western North America,* ed. Earl H. Swanson (Pocatello: Idaho State University Press, 1970), 201–231; and Brigham D. Madsen, *The Bannock of Idaho* (Caldwell, Idaho: Caxton Printers, 1958).

28. David Dominick, "The Sheepeaters," *Annals of Wyoming* 36 (1964): 131–168; Åke Hultkrantz, "The Shoshones in the Rocky Mountain Area," *Annals of Wyoming* 29 (1957): 125–149; Hultkrantz, "The Indians in Yellowstone Park," 19–41; Murphy and Murphy, "Shoshone-Bannock Subsistence and Society," 322–323; and Loendorf, "Prehistory of the Sheep Eater Indians."

29. Osborne Russell, *Osborne Russell's Journal of a Trapper,* ed. Aubrey L. Haines (Lincoln: University of Nebraska Press, 1965) 26–27. Also see Sven Liljeblad *The Idaho Indians in Transition, 1805–1960* (Pocatello: Idaho State University Museum, 1972), 38.

30. In the late 1870s, evidence of abandoned Sheep Eater winter camps could be found "in nearly all of the sheltered glens and valleys of the Park": Philertus W. Norris, *Annual Report of the Superintendent of the Yellowstone National Park to the Secretary of the Interior for the Year 1880* (Washington, D.C.: Government Printing Office, 1880), 35.

31. Dominick, "The Sheepeaters," 154–156; Hultkrantz, "The Indians in Yellowstone Park," 139.

32. For a rendering of the Crow migration story, see Joe Medicine Crow, *From the Heart of Crow Country: The Crow Indians' Own Stories* (New York: Crown, 1992), 16–24. Robert Lowie suggested that Crow use of the Yellowstone area may have begun more than four hundred years ago: *The Crow Indians* (Lincoln: University of Nebraska Press, 1983), 3.

33. For a specific reference to the Crow as the "Rocky Mountain Indians" see François Antoine Larocque's "Yellowstone Journal," in *Early Fur Trade on the Northern Plains: Canadian Traders among the Mandan and Hidatsa Indians, 1738–1818,* ed. W. Raymond Wood and Thomas D. Thiessen (Norman: University of Oklahoma Press, 1985), 165.

34. Lloyd "Mick" Old Coyote, interview with the author, Crow Agency, Mont., July 8, 1994.

35. "Lieut. James H. Bradley Manuscripts," *Contributions to the Montana Historical Society* 9 (1923): 306–307.

36. The Crow's devotion to their homeland is legendary. See Medicine Crow, *From the Heart of Crow Country;* and Eloise Whitebear Pease, *Crow Tribal Treaty Centennial Issue* (Crow Agency, Mont.: Crow Tribe, 1968).

37. Not surprisingly, the mountains that run along the eastern boundary of Yellowstone National Park are called the Absaroka Range. *Absaroka,* which roughly translates as "large beaked bird" or "people of the large beaked bird," is a common rendering of the Crow word for themselves. Variously known as the Yellowstone Range, Yellowstone Mountains, Upper Yellowstone Mountains, Great Yellowstone Range, and Stinkingwater Mountains, the Absaroka Range received its current name in 1885 from the U.S. Geological Survey. Lee H. Whittlesey, *Yellowstone Place Names* (Helena: Montana Historical Society Press, 1988), 13.

38. For an overview of the fur trade era in Yellowstone and various trappers' misadventures with the Blackfeet, see Haines, *Yellowstone National Park,* 6–41. For Blackfeet expansion toward the headwaters of the Missouri and conflict with the Shoshone, see John C. Ewers, *The Blackfeet, Raiders on the Northwestern Plains* (Norman: University of Oklahoma Press, 1958), 15–30.

39. For brief discussions of plains and intermontane groups' use of the Yellowstone region in the first half of the nineteenth century, see Haines, *The Yellowstone Story: A History of Our First National Park,* rev. ed., 2 vols. (Niwot, Colo.: Yellowstone Association for Natural Science, History & Education, in cooperation with University Press of Colorado, 1996), 1: 27–30; Hultkrantz, "The Indians in Yellowstone Park," 142–143; and Janetski, *Indians of Yellowstone,* 29–32.

40. Shimkin, "Wind River Shoshone Ethnogeography," 254–255; Frederic E. Hoxie, *Parading Through History: The Making of the Crow Nation in America, 1805–1935* (Cambridge: Cambridge University Press, 1995) 130–136.

41. For analysis of the importance of elk as a food source, see Richard E. McCabe, "Elk and Indians: Historical Values and Perspectives," in *Elk of North America: Ecology and Management,* ed. Jack Ward Thomas and Dale E. Toweill (Harrisburg, Pa.: Stackpole Books, 1982), 87–93. The near extinction of the buffalo in the early 1880s led to its almost wholesale replacement by elk among the Wind River Shoshone: S. R. Martin, "Report of Shoshone Agency, Wyoming," *ARCIA, 1885,* 212.

42. For an overview of the problems attendant with white migration and settlement in southeastern Idaho in the 1840s through 1860s, see Brigham D. Madsen, *The Shoshoni Frontier and the Bear River Massacre* (Salt Lake City: University of Utah Press, 1985).

43. Raymond J. DeMallie, "Touching the Pen: Plains Indian Treaty Councils in Ethnohistorical Perspective," in *Ethnicity on the Great Plains,* ed. Frederick C. Luebke (Lincoln: University of Nebraska Press, 1980), 38–51; Richard White, "The Winning of the West: The Expansion of the Western Sioux in the Eighteenth and Nineteenth Centuries," *Journal of American History* 65 (1978): 319–343.

44. Charles J. Kappler, comp., *Indian Affairs: Laws and Treaties,* 5 vols. (Washington: Government Printing Office, 1904–1941), 2: 594–596;

45. The 1868 Fort Laramie and Fort Bridger Treaties are reprinted in ibid., 2: 1008–1011 and 1021–1024. The executive order of July 8, 1869, establishing the Fort Hall Reservation is reprinted in ibid., 1: 838–839.

46. See W. H. Danilson, "Report of Fort Hall Agency," and Harrison Fuller, "Report of Lemhi Agency," *ARCIA, 1876,* 258–260 and 44–45.

47. The "Treaty with the Shoshones, Bannacks, and Sheepeaters" is reprinted in Kappler, *Indian Affairs,* 4: 707–708.

48. Philetus W. Norris, *Report upon the Yellowstone National Park to the Secretary of the Interior for the Year 1879* (Washington, D.C.: Government Printing Office, 1880), 10.

49. John Wright, "Report of Lemhi Agency," *ARCIA, 1878* and *ARCIA, 1879,* 51–52 and 54–55. Other Sheep Eater groups also made occasional use of the Fort Hall and Wind River reservations; see Dominick, "The Sheepeaters."

50. For recent comments on the end of native use of the park area in 1871, see Haines, *Yellowstone Story,* 1: 29; and Whittlesey, *Yellowstone Place Names,* 138.

51. The Works Progress Administration carried out a nationwide program of oral histories in the 1930s. Those conducted in Park County, Montana, which borders Yellowstone National Park to the north, contain numerous observations about Crow in the area through the late 1870s: "Works Progress Administration Questionnaires for Park County, Montana," Merrill G. Burlingame Special Collections, Montana State University Library, Bozeman, Mont.

52. John P. C. Shanks, T. W. Bennet, and Henry W. Reed, "Report of Special Commissioners to Investigate and Report upon Indian Affairs in the Territory of Idaho, Adjacent Territories," *ARCIA, 1873,* 157–159; James Irwin, "Report of Shoshone and Bannack Agency, Wyoming," *ARCIA, 1873,* 244–245; W. H. Danilson, "Report of Fort Hall Agency," *ARCIA, 1875* and *ARCIA, 1877,* 258–260 and 78–79; Fuller, "Report of Lemhi Agency," *ARCIA, 1876,* 44–45; and Wright, "Report of the Lemhi Agency," *ARCIA, 1878,* 51–52.

53. William A. Jones, *Report upon the Reconnaissance of Northwestern Wyoming, Including Yellowstone National Park, Made in the Summer of 1873* (Washington, D.C.: Government Printing Office, 1875), 19, 22, 36, 54–55; Norris, "Journals" letter no. 21, August 30, 1875; William S. Brackett, "Indian Remains on the Upper Yellowstone," *The American Field* 39 (1893): 127–128; Phillip H. Sheridan, *Report of Lieut. General P. H. Sheridan, dated September 20, 1881, of his expedition through the Big Horn Mountains, Yellowstone National Park, Etc.* (Washington, D.C.: Government Printing Office, 1882), 8–10; Walter Scribner Schuyler, "Journal of Surveying Party to Yellowstone, July 29–September 10, 1883," Walter Scribner Schuyler Collection, Henry E. Huntington Library, San Marino, Calif., 24–25, and attached map indicating Indian camps within park; E. S. Topping, *The Chronicles of the Yellowstone* (St. Paul: Pioneer Press, 1888), 98–99; and William A. Baillie-Grohman, *Camps in the Rockies* (New York: Charles Scribner's Sons, 1882), 263–277.

54. Haines, *Yellowstone Story,* 2: 59–60, 483–484; Steven W. Chadde and Charles E. Kay, "Tall Willow Communities on Yellowstone's Northern Range: A Test of the 'Natural Regulation' Paradigm," in *The Greater Yellowstone Ecosystem: Redefining America's Wilderness Heritage,* ed. Robert B. Keiter and Mark S. Boyce (New Haven: Yale University Press, 1991), 231–264.

55. The commissioners had originally come to negotiate the purchase of the entire reservation and move the Crow north of the Yellowstone River to the Judith Basin. The Crow refused to "touch the pen" on the final agreement, and the Senate never ratified it. See "Report of the Commission to Negotiate with the Crow Tribe of Indians," *ARCIA, 1873,* 114–143. Blackfoot died in the disputed area in the summer of 1879 while leading a hunting party toward the eastern boundary of the national park; Augustus R. Keller to Commissioner of Indian Affairs (CIA), July 3, 1879, box 9, folder 2, "Press Copies of Letters Sent to the Bureau of Indian Affairs and Other Correspondents," Records of the Crow Indian Agency, Bureau of Indian Affairs, Record Group 75; National Archives–Pacific Northwest Region, Seattle (hereafter NA-PNR RG 75, BIA, "Press Copies"). Also see Medicine Crow, *Heart of Crow Country,* 38–39.

56. This was the camp of Iron Bull and Blackfoot, the two most important leaders of the Crow. See Lt. G. C. Doane, "U.S. Interior Department, Office of Indian Affairs Report," February 19, 1874, doc. no. SC 889, Montana Historical Society Library and Archives, Helena, Mont. (hereafter MHS).

57. Quotation from George Frost to CIA, September 12, 1877, "Crow Indian Agency Records," box 1, doc. no. MC87, MHS. Also, Dexter E. Clapp, "Report of Crow Agency," *ARCIA, 1875,* 301; Keller to CIA, June 19, 1879, NA-PNR RG 75, BIA, "Press Copies," box 9, folder 2; Keller to CIA, September 3, 1881, NA-PNR RG 75, BIA, "Press Copies," box 9, folder 6; Keller, "Report of Crow Agency," *ARCIA, 1879,* 91–94; and Schuyler, "Journal of Surveying Party," 24.

58. Keller to CIA, March 7, 1881, NA-PNR RG 75, BIA, "Press Copies," box 9, folder 5; Schuyler, "Journal of Surveying Party," 24; and Sheridan, *Report of Lieut. General P. H. Sheridan,* 8–9.

59. Charles C. Bradley Jr., "After the Buffalo Days: An Account of the First Years of Reservation Life for the Crow Indians, Based on Official Government Documents from 1880 to 1904 A.D.," 1971, unpublished manuscript in the Little Big Horn College Library, Crow Agency, Mont.; Hoxie, *Parading Through History,* 96–166. Six band chiefs ceded the western portion of the reservation in 1880. The agreement was not ratified by the U.S. Senate until 1882, and the tribe did not move to their new agency until 1884. Two versions of a tragic story about Crow warriors who took a raft over Yellowstone Falls and committed suicide instead of surrendering to pursuing troops of the U.S. Army suggests how painful the loss of the Yellowstone area must have been for some individuals; see Charles M. Skinner, *Myths and Legends of Our Land,* 2 vols. (Philadelphia: Lippincott, 1896), 2: 204–206; and Clark, *Indian Legends,* 323–324.

60. E. S. Topping, *The Chronicles of the Yellowstone* (St. Paul: Pioneer Press, 1888), 6; and Hultkrantz, "The Indians in Yellowstone Park," 145.

61. Wright, "Report of the Lemhi Agency," *ARCIA 1879,* 55; Irwin, "Report of Shoshone Agency, Wyoming," *ARCIA, 1883,* 313–314; and Robert Woodbridge, "Report of Lemhi Agency," *ARCIA, 1885,* 68.

62. In his essay "Aboriginal Overkill and Native Burning," Charles Kay argues that Indian overhunting throughout the nineteenth century had made Yellowstone National Park a "game-poor" region at the time of its establishment in 1872. He bases much of his argument on contradictory evidence, however. At one point he cites the Washburn expedition's inability to keep itself supplied with sufficient meat without acknowledging the frequent signs of large herds of elk and other game animals that Langford and Doane commented on. Unsuccessful hunting did not result from a dearth of game animals but instead reflected the objectives of the expedition. Because the Washburn party sought out geysers and mountain peaks and did not pause long enough to adequately stalk game, they had very few opportunities to bring down anything larger than a grouse. For observations on the park's large game populations in the 1870s, see Norris, "Journals" letter no. 22, September 1, 1875; and W. E. Strong, *A Trip to the Yellowstone National Park in July, August, and September, 1875* (Norman: University of Oklahoma Press, 1968), 47, 104–107.

63. Jones, *Reconnaissance of Northwestern Wyoming,* 19, 54–55.

64. Strong, *Trip to the Yellowstone,* 105.

Notes to Chapter 4

1. George W. Wingate, *Through the Yellowstone Park on Horseback* (New York: O. Judd, 1886), 36.

2. Aubrey L. Haines estimates that only three hundred visitors came to the new national park in 1872, and probably no more than five hundred in the years prior to 1877; see

his *The Yellowstone Story: A History of Our First National Park,* rev. ed., 2 vols. (Niwot, Colo.: Yellowstone Association for Natural Science, History & Education, in cooperation with University Press of Colorado, 1996), 1:196.

3. Hiram Martin Chittenden, *The Yellowstone National Park, Historical and Descriptive* (Cincinnati: Robert Clarke, 1895), 79. Alfred Runte makes this observation a cornerstone of his own argument about early national park history; see his *National Parks: The American Experience,* 3d ed. (Lincoln: University of Nebraska Press, 1997), 35–40, 43–44, 54–55.

4. Lucullus Virgil McWhorter, *Yellow Wolf: His Own Story* (Caldwell, Idaho: Caxton Printers, 1940), 130.

5. The early fame of the Nez Perce War has inspired a great deal of scholarship; see, for example, Merril D. Beal, *"I Will Fight No More Forever": Chief Joseph and the Nez Perce War* (Seattle: University of Washington Press, 1963); Mark H. Brown, *The Flight of the Nez Perce* (New York: G. P. Putnam Sons, 1967); and Bruce Hampton, *Children of Grace: The Nez Perce War of 1877* (New York: Henry Holt, 1994). For an excellent analysis of Nez Perce movements in the park, see William L. Lang, "Where Did the Nez Perce Go?" *Montana, the Magazine of Western History* 40 (1990): 14–29.

6. Philetus W. Norris, *Report upon the Yellowstone National Park to the Secretary of the Interior for the Year 1878* (Washington, D.C.: Government Printing Office, 1879), 979.

7. Ibid., 981. For a general account of the Bannock War, see Brigham D. Madsen, *The Northern Shoshoni* (Caldwell, Idaho: Caxton Printers, 1980), 75–89.

8. For an account of the so-called Sheep Eater War, see Madsen, *The Lemhi: Sacajawea's People* (Caldwell, Idaho: Caxton Printers, 1979), 102–105; and Sven Liljeblad, T*he Idaho Indians in Transition, 1805–1960* (Pocatello: Idaho State University Museum, 1972), 38–39.

9. Quotes are from P. H. Conger, *Report of the Superintendent of the Yellowstone National Park, 1883* (Washington, D.C.: Government Printing Office, 1883), 5. Also see David G. Battle and Erwin N. Thompson, *Yellowstone National Park: Fort Yellowstone Historic Structure Report* ([Washington: D.C.]: National Park Service, 1972).

10. Norris, *Report upon the Yellowstone National Park to the Secretary of the Interior for the Year 1879* (Washington, D.C.: Government Printing Office, 1880), 10, 23, 31; Norris, *Annual Report of the Superintendent of the Yellowstone National Park to the Secretary of the Interior for the Year 1880* (Washington, D.C.: Government Printing Office, 1880), 7, 17, 23.

11. Norris, *Report for 1879,* 10. For Norris's ongoing concerns about augmenting tourist facilities and attracting concessionaires to the park, see his *Report for 1878,* 985–987.

12. E. S. Topping, *The Chronicles of the Yellowstone* (St. Paul: Pioneer Press, 1888), 6; Åke Hultkrantz, "The Indians in Yellowstone Park," *Annals of Wyoming* 29 (1957): 145.

13. Norris, *Annual Report for 1880,* 70.

14. Norris, *Report for 1879,* 12.

15. Crow resistance to the 1880 land cession agreement is detailed in C. H. Barstow, "Report of Council Held with the Crows at Crow Agency, M.T., June 12th, 1880," Record Group 75, Records of the Bureau of Indian Affairs, "Letters Received by the Office of Indian Affairs, 1824–1881," microfilm series M234, roll 516, frames 586–599 (hereafter RG 75 LR 1824–1881).

16. Plenty Coups is quoted in 51st Cong., 2d sess., 1890, S. Doc. 43, serial 2818, 11–12. For an overview of the 1880 land cession agreement and the relocation of the Crow agency, see Frederick E. Hoxie, *Parading Through History: The Making of the Crow Nation in America,* 1805–1935 (Cambridge: Cambridge University Press, 1995), 106–166.

17. Quotes are from Norris, *Annual Report for 1880,* 3. For his efforts to influence the Washington negotiations, see Norris to Commissioner of Indian Affairs (CIA), April 26, 1880, RG 75 LR 1824–1881. M234, roll 352, frame 322. For his meetings with the Shoshone and Bannock in Idaho, see Norris to Carl Schurz, June 21, 1880, and Norris to Harry Yount, June 21, 1880; both in Record Group 48, "Records of the Office of the Secretary

of the Interior Relating to Yellowstone National Park," microfilm series M62, roll 1, frames 288–289 (hereafter RG 48 YNP).

18. Norris, *Fifth Annual Report of the Superintendent of the Yellowstone National Park to the Secretary of the Interior* (Washington, D.C.: Government Printing Office, 1881), 45. The final agreements with the Crow, Shoshone, Bannock, and Sheep Eater are reprinted in *Annual Report of the Commissioner of Indian Affairs, 1880* (Washington, D.C.: Government Printing Office, 1881), 277–279 (hereafter *ARCIA*).

19. Norris, *Report upon the Yellowstone National Park to the Secretary of the Interior* (Washington, D.C.: Government Printing Office, 1877), 10.

20. Ibid., 11; emphasis in original.

21. Norris, *Fifth Annual Report,* 45.

22. For the comments of agency personnel regarding off-reservation travel, see A. L. Cook, "Report of Fort Hall Agency," *ARCIA, 1882,* 49–51; John Harries, "Report of Lemhi Agency," *ARCIA, 1883,* 55–56; Charles Hatton, "Report of Shoshone and Bannock Agency, Wyoming," *ARCIA, 1881,* 182–185; James Irwin, "Report of Shoshone Agency, Wyoming," *ARCIA, 1883,* 313–314; Woodbridge, "Report of Lemhi Agency," *ARCIA, 1885,* 63–66.

23. Conger, *Report of the Superintendent,* 5.

24. Wingate, *Through the Yellowstone,* 36.

25. Some of the more popular guidebooks in the early 1880s were Herman Haupt, *The Yellowstone National Park* (New York: J. M. Stoddart, 1883); George L. Henderson, *Yellowstone Park Manual and Guide* (Mammoth Hot Springs, Yellowstone National Park: [G. L. Henderson], 1885); Henry J. Winser, *The Yellowstone National Park: A Manual for Tourists* (New York: G. P. Putnam's Sons, 1883); and W. W. Wylie, *Yellowstone National Park; of the Great American Wonderland* (Kansas City, Mo.: Ramsey, Millett & Hudson, 1882). Published accounts of early vacations in Yellowstone include W. H. Dudley, *The National Park from the Hurricane Deck of a Cayuse, or, the Liederkrantz Expedition to Geyserland* (Butte City, Mont.: F. Loeber, 1886); Almon Gunnison, *Rambles Overland: A Trip Across the Continent* (Boston: Universalist Press, 1884); J. E. Williams, *Through Yellowstone National Park: Vacation Notes, Summer of 1888* (Amherst, Mass.: Amherst Record, 1889); Mary Bradshaw Richards, *Camping Out in the Yellowstone, 1882,* ed. William W. Slaughter (Salt Lake City: University of Utah Press, 1994); Margaret Andrews Allen, "A Family Camp in Yellowstone Park," *Outing* (1885): 157–159. Numerous tourist diaries are housed in the Yellowstone National Park Archives at Mammoth Hot Springs, Yellowstone National Park; the Henry E. Huntington Library, San Marino, Calif.; the Merrill G. Burlingame Special Collections, Montana State University Library, Bozeman, Mont.; and the Montana Historical Society, Helena.

26. For a discussion of the railroad proposals and the "Yellowstone war," see Haines, *Yellowstone Story,* 2: 30–38 and 54–59.

27. Vest quoted in ibid., 2: 35.

28. O'Neil quoted in ibid., 2: 36. For President Arthur's trip to Yellowstone, see *Journey Through the Yellowstone National Park and Northwestern Wyoming, 1883* (n.p: n.d.), copy in the Yellowstone National Park Archives, Mammoth, Wyo., Yellowstone National Park (hereafter YNPA).

29. Haines, *Yellowstone Story,* 2: 38–48; Runte, *National Parks,* 54–55; Roderick Nash, *Wilderness and the American Mind,* 3d ed. (New Haven: Yale University Press, 1982), 115–116.

30. Quote is from *U.S. Statutes at Large* 392, (1872): 32–33. Historian Richard West Sellars makes a similar argument about the shift from scenic interests to concerns about wildlife conservation at this time; see his *Preserving Nature in the National Parks: A History* (New Haven: Yale University Press, 1997), 24–25.

31. Lucius Q. C. Lamar to Senator Charles F. Manderson, April 22, 1886, RG 48, YNP M62, roll 6, frames 57–82. Lamar's twenty-six-page letter to Manderson not only summa-

rized the many debates surrounding the management of Yellowstone National Park but also served as the basis for his annual report; see *Annual Report of the Secretary of the Interior for 1886* (Washington, D.C.: Government Printing Office, 1886).

32. For early complaints about commercial hunters in Yellowstone, see W. E. Strong, *A Trip to the Yellowstone National Park in July, August, and September, 1875* (Norman: University of Oklahoma Press, 1968), 105; and Norris, *Annual Report for 1877*, 13.

33. William Ludlow, *Report of a Reconnaissance from Carroll, Montana Territory, on the Upper Missouri, to the Yellowstone National Park, and Return, Made in the Summer of 1875* (Washington, D.C.: Government Printing Office, 1876), 36–37. George Bird Grinnell, who served as the geologist on Ludlow's reconnaissance of Wyoming Territory and the national park, seconded the captain's recommendation that only a military police force "might . . . prevent the reckless destruction of the animals"; Grinnell, "Letter of Transmittal," in Ludlow, *Report of a Reconnaissance.*

34. Lamar to Manderson, April 22, 1886.

35. Philip H. Sheridan, *Report of Lieut. General P. H. Sheridan, Dated September 20, 1881, of His Expedition Through the Big Horn Mountains, Yellowstone National Park, Etc.* (Washington, D.C.: Government Printing Office, 1882), 9; and *Report of an Exploration of Parts of Wyoming, Idaho, and Montana, in August and September, 1882* (Washington, D.C.: Government printing Office, 1882), 17–18.

36. Quote is from *U.S. Statutes at Large* 2317 (1883): 626. Also see Haines, *Yellowstone Story*, 1: 267–269, 2: 59; and Louis C. Cramton, *Early History of Yellowstone National Park and Its Relation to National Park Policies* (Washington, D.C.: Government Printing Office, 1932), 41–43. For a detailed study of military management of Yellowstone National Park at this time, see Karl Jacoby, "The Recreation of Nature: A Social and Environmental History of American Conservation, 1872–1919" (Ph.D. diss., Yale University, 1997), 201–250; also see H. Duane Hampton, *How the U.S. Cavalry Saved Our National Parks* (Bloomington: Indiana University Press, 1971).

37. [Ezra A. Hayt], *ARCIA, 1879*, xxvi. These complaints were specifically leveled at a group of Mountain Ute who had left their reservation to return to Middle Park in Colorado. Many of these same people had no doubt encountered Samuel Bowles and his party in the summer of 1868.

38. James B. Trefethen, *An American Crusade for Wildlife* (New York: Winchester Press and the Boone and Crockett Club, 1975), 79.

39. For a report of Indian "incursions" before Harris's arrival, see H. L. Muldrow to Superintendent of Yellowstone National Park, June 5, 1886, Army Records, document 108, YNPA (hereafter AR YNPA, plus document number). Harris officially took charge of Yellowstone on August 21, 1886, and made his first complaint about Indians along the park's western boundary on August 26; Harris to Muldrow (telegram), August 26, 1886, Army Records, "Letters Sent," vol. 1, YNPA (hereafter AR YNPA "Letters Sent," plus volume number). Quotes are from Harris, *Annual Report of the Superintendent of the Yellowstone National Park, 1886* (Washington, D.C.: Government Printing Office, 1886), 7.

40. Harris to Muldrow, February 12, 1889; reprinted in Harris, *Report of the Superintendent of the Yellowstone National Park, 1889* (Washington, D.C.: Government Printing Office, 1889), 15–16.

41. Harris to Muldrow, August 22, 1887, National Archives, record group 75, Records of the Bureau of Indian Affairs, letters received, 1881–1907, box 416, doc. 22870 (hereafter NA RG 75, BIA LR 1881–1907); Harris to Peter Gallagher, October 2, 1888, AR YNPA, 377. While most hunting parties numbered between two and three dozen individuals, Harris estimated that a group of one hundred Bannock approached Yellowstone in the summer of 1888; Harris to Muldrow, August 24, 1888, AR YNPA "Letters Sent," vol. 2.

42. Harris to Muldrow, May 4, 1888, NA RG 75, BIA LR 1881–1907, 462/12385.

43. Quotes are from a restatement of these policies in CIA to Secretary of the Interior, January 19, 1889, AR YNPA, 375. Also see A. B. Upshaw to Harris, May 23, 1888, AR YNPA, 874.

44. J. M. Needham to CIA, May 31, 1888, NA RG 75, BIA LR 1881–1907, 467/14646; and Gallagher to CIA, June 16 and August 13, 1888, NA RG 75, BIA LR 1881–1907, 469/15733 and 478/20827.

45. Harris's report to the secretary of the Interior is included in *Report of the Superintendent, 1889,* 13–16. For a sample of his many letters to the CIA, the Secretary of the Interior, and agents Gallager and Needham, see CIA to Secretary of the Interior and enclosures, January 19, 1889, AR YNPA, 375–385. Harris's successor would later refer to these many letters as a "great deal of unpleasant correspondence"; Captain F. A. Boutelle to U.S. Indian Agent, Lemhi Agency, June 11, 1889, AR YNPA "Letters Sent," vol. 2.

46. Quote is from the constitution of the Boone and Crockett Club as reprinted in Grinnell, ed., *Hunting at High Altitudes* (New York: Harper & Brothers, 1913), 436–437. Also see John F. Reiger, *American Sportsmen and the Origins of Conservation* (New York: Winchester Press, 1975), 101–109.

47. George Bird Grinnell, "Indian Marauders," "A Case for Prompt Action," and "Indians and the National Park," *Forest and Stream* 32 (April 4, April 11, and May 2, 1889), 209, 233–235, and 296; also see Anonymous, "Protect the National Park," *Frank Leslie's Illustrated Newspaper* 68 (April 27, 1889), 182.

48. Boutelle to U.S. Indian Agent, Lemhi Agency, June 11, 1889, AR YNPA, 438; Needham to Boutelle, n.d., 1889, AR YNPA, 1206; and E. Nasholds to Boutelle, August 30, 1890, AR YNPA, 1057. Also see S. G. Fisher to CIA, September 3, 1890, NA RG 75, BIA LR 1881–1907, 658/27771; and Captain George S. Anderson to Secretary of the Interior, February 7, 1893, AR YNPA "Letters Sent," vol. 4.

49. See, for example, William T. Hornaday, "The Extermination of the American Bison, with a Sketch of Its Discovery and Life History," *Annual Report of the Smithsonian Institution, 1887,* vol. 2. (Washington, D.C.: Government Printing Office, 1889); and George Bird Grinnell, *The Last of the Buffalo* (New York: Forest & Stream Publishing, 1894).

50. Haines, *Yellowstone Story,* 2: 60–68. For an excellent study of hunting and park policy at this time, see Jacoby, "The Recreation of Nature," 251–273. Quote is from *U.S. Statutes at Large* 6442, (1894): 73.

51. Quote is from "Petition of the residents of Uinta and Fremont Counties, State of Wyoming," sent to the Superintendent of Yellowstone National Park in September 1893, AR YNPA, 816. For a statement of support from Acting Superintendent George S. Anderson, see Anderson to Ira Dodge, October 4, 1893, AR YNPA "Letters Sent," vol. 4. At the time, Uinta County embraced all of what are now the counties of Teton, Lincoln, and Uinta.

52. *ARCIA, 1894,* 66–67, 75.

53. Ibid., 66–68, 75–77. At the time, the present town of Jackson, Wyoming, was known as Marysvale.

54. Ibid., 76. For an excellent analysis of the many historical issues that shaped the struggle for control of the elk in Jackson Hole, see Louis S. Warren, *The Hunter's Game: Poachers and Conservationists in Twentieth-Century America* (New Haven: Yale University Press, 1997), 1–20.

55. *ARCIA, 1894,* 67–80. Se-we-a-gat is also referred to as Tanega or Timega in several reports to the commissioner of Indian Affairs.

56. Ibid., 77.

57. Ibid., 63–64.

58. Ibid., 65–66; Brigham D. Madsen, *The Bannock of Idaho* (Caldwell, Idaho: Caxton Printers, 1958), 261–262.

59. *ARCIA, 1896,* 56–60. For a discussion of Marshall's opinion in *Worcester v. Georgia* and its importance for later interpretations of federal Indian law, see Charles F. Wilkinson, *American Indians, Time, and the Law* (New Haven: Yale University Press, 1987), 30–37, 55–62.

60. *ARCIA, 1896,* 61. All subsequent quotes of the court's decision are from Justice Edward Douglas White's opinion for the majority in *Ward v. Race Horse,* 163 U.S. 504 (1896); reprinted in *ARCIA, 1896,* 60–66.

61. Ibid., 60–62. For an interpretation of the court's decision in light of late-nineteenth-century ideas about the frontier in American history, see Warren, *Hunter's Game,* 4–5.

62. *ARCIA, 1896,* 63.

63. Walter F. Pratt Jr., "Plessy v. Ferguson," in *The Oxford Companion to the Supreme Court of the United States,* ed. Kermit L. Hall (New York: Oxford University Press, 1992), 637–638. *Ward v. Race Horse* has not received the scholarly attention it deserves, but some of the later implications of the case are discussed in Raymond Cross and Elizabeth Brenneman, "Devils Tower at the Crossroads: The National Park Service and the Preservation of Native American Cultural Resources in the 21st Century," *Public Land & Resources Review* 18 (1997): 5–45. Also see Brian Czech, "*Ward vs Racehorse*—Supreme Court as Obviator?" *Journal of the West* 35 (1996): 61–69.

64. The Bannock eventually received $75,000 from the government in 1900 for the loss of their hunting rights.

65. Sidney L. Harring, *Crow Dog's Case: American Indian Sovereignty, Tribal Law, and United States Law in the Nineteenth Century* (Cambridge: Cambridge University Press, 1994), 204–206.

66. "Diary of Yellowstone Park Scouts, Winter Season of 1897–98," 67, YNPA.

67. Captain Wilbur E. Wilder to E. C. Waters, March 29, 1899; E. A. Hitchcock to Wilder, April 15, 1899; both in AR YNPA "Letters Sent," vol. 8.

68. One early park concessionaire referred to Yellowstone as the "Eden of America"; advertisement for Bassett Bros. in the July 30, 1882, issue of the *Salt Lake City Tribune;* reprinted in Richards, *Camping Out in the Yellowstone,* 13.

69. Haines, *Yellowstone Story,* 2: 483–484.

Notes to Chapter 5

1. George Bird Grinnell, *Blackfoot Lodge Tales* (New York: Scribner's, 1892), 142–143.

2. Quotation from advertisement in the *Saturday Evening Post,* May 15, 1926, reproduced in Great Northern Railway Company, Advertising and Publicity Department, "Magazine and Newspaper Advertisements, 1884–1970," microfilm edition, volume 1, roll 1, frame 40, Minnesota Historical Society, Minneapolis.

3. See Brian Reeves and Sandy Peacock, "'Our Mountains Are Our Pillows': An Ethnographic Overview of Glacier National Park," 2 vols., draft submitted to the National Park Service, Rocky Mountain Region, Denver, May 1995, xi–xiv. For a comparative study of these three groups' relations with the Canadian and U.S. governments, see Hana Samek, *The Blackfoot Confederacy, 1880–1920: A Comparative Study of Canadian and U.S. Indian Policy* (Albuquerque: University of New Mexico Press, 1987).

4. Reeves and Peacock, "'Our Mountains Are Our Pillows,'" 1: 7–69; James Sheire, *Glacier National Park: Historic Resource Study* (Washington, D.C.: U.S. Department of the Interior, National Park Service, 1970), 3–47; and C. W. Buchholtz, *Man in Glacier* (West Glacier, Mont.: Glacier Natural History Association, 1976), 1–25. The rights reserved to the Blackfeet were last recognized in an 1895 agreement between the tribe and the United States. Although the government had ceased making treaties with Indian tribes in 1871, the agreement reserved certain rights that were first delineated in two previous treaties.

5. The Blackfeet have a very rich oral tradition, and many of their stories concern the Glacier region. For the most accessible written collection of stories, see Percy Bullchild's *The Sun Came Down: The History of the World as My Blackfeet Elders Told It* (New York: Harper and Row, 1985). As with all oral traditions, some variation occurs over time and between different storytellers, and much is altered in the process of transcription and translation. Nevertheless, the Napi stories have remained constant in nearly all of their geographic particulars since the first recording by an outsider in 1810. See David Thompson, *David Thompson's Narrative of His Explorations in Western America, 1784–1812,* ed. J. B. Tyrrell (Toronto: Champlain Society, 1916), 326–334; John Mason Brown, "Traditions of the Blackfeet," *Galaxy* January 15, 1867, 157–164; Grinnell, *Blackfoot Lodge Tales,* 137–176; Clark Wissler and D. C. Duvall, "Mythology of the Blackfoot Indians," in *Anthropological Papers of the American Museum of Natural History,* vol. 2, part 1 (New York: American Museum of Natural History, 1908), 1–163; Walter McClintock, *The Old North Trail or Life, Legends and Religion of the Blackfeet Indians* (London: Macmillan, 1910); James Willard Schultz, *Blackfeet Tales of Glacier National Park* (Boston: Houghton Mifflin, 1916); and Jack Holterman, "Seven Blackfeet Stories," *Indian Historian* 3 (1970): 39–43.

6. For accounts of the origins of the Beaver Pipe Bundle and the sacred tobacco, see Grinnell, *Blackfoot Lodge Tales,* 117–124; Wissler and Duvall, "Mythology of the Blackfoot," 74–80; McClintock, *Old North Trail,* 104–112; Schultz, *Blackfeet Tales,* 216–224; and Bullchild, *The Sun Came Down,* 290–324. Wissler and Duvall cite Upper St. Mary's Lake (South Big Inside Lake) as the original locale of the bundle. Bullchild places it at Waterton Lake (North Big Inside Lake). Upper St. Mary's Lake is within Glacier National Park; Waterton Lake is on the boundary between Glacier and Waterton Lakes National Park in Alberta. These two national parks comprise the Glacier-Waterton International Peace Park. Reeves and Peacock estimate that the original bundle came to the Blackfeet from the Glacier area more than a thousand years ago: "'Our Mountains Are Our Pillows,'" 1: 160.

7. Schultz, *Blackfeet Tales,* 158–182.

8. Clark Wissler, "Ceremonial Bundles of the Blackfoot Indians," *Anthropological Papers of the American Museum of Natural History,* vol. 7, part 2 (New York: American Museum of Natural History, 1912); McClintock, *Old North Trail,* 424–426; and Schultz, *Blackfeet Tales,* 23–42. As with ongoing concerns about native use of the national park, many of the spiritual beliefs discussed here remain an important part of contemporary Blackfeet life.

9. For a broader discussion of the sacred importance of the Glacier area for the Blackfeet, see J. Hansford C. Vest, "Traditional Blackfeet Religion and the Sacred Badger Two-Medicine Wildlands," *Journal of Law and Religion* 6 (1988): 455–489; and Bob Yetter, *Badger-Two Medicine, The Last Stronghold: Sacred Land of the Grizzly, Wolf, and Blackfeet Indian* (Missoula, Mont.: Badger Chapter of Glacier–Two Medicine Alliance, 1992). By far the most comprehensive study on the historical and contemporary use of the Glacier region by the Blackfeet is Reeves and Peacock, "'Our Mountains are Our Pillows.'" Their work has informed much of the foregoing discussion, and I am especially grateful to the authors for allowing me to see a finished draft of this work and to David Ruppert of the National Park Service for providing me with a copy.

10. Clark Wissler, "Material Culture of the Blackfoot Indians," in *Anthropological Papers of the American Museum of Natural History,* vol. 2, part 1 (New York: American Museum of Natural History, 1908) 11–20; Reeves and Peacock, "'Our Mountains Are Our Pillows,'" 1: 70–82. Reeves convincingly demonstrates the errors of theories about Blackfeet migration to the Rocky Mountain region in historical times. Noting the absence of a migration story within Blackfeet oral tradition, he utilizes linguistic, archaeological, and early archival evidence to refute more popular theories about the Blackfeet's leaving the Great Lakes region sometime in the late seventeenth century.

11. Vernon Bailey, *Wild Animals of Glacier National Park* (Washington, D.C.: Govern-

ment Printing Office, 1918); McClintock, "Appendix: Medicinal and Useful Plants of the Blackfoot Indians," in *Old North Trail,* 524–531; Alex Johnston, *Plants and the Blackfoot* (Lethbridge, Alberta: Lethbridge Historical Society and the Historical Society of Alberta, 1987); and Karen Peacock, "Appendix II: Ethnobotanical Plant Descriptions," in Reeves and Peacock, "'Our Mountains Are Our Pillows,'" 2: 334–472.

12. John C. Ewers, "The Horse in Blackfoot Indian Culture, with Comparative Material from Other Western Tribes," *Bureau of American Ethnology, Bulletin 159* (Washington, D.C.: Bureau of American Ethnology, 1955); Oscar Lewis, "The Effects of White Contact upon Blackfoot Culture with Special Reference to the Role of the Fur Trade" (Ph.D. diss., Columbia University, 1942), copy in the Henry E. Huntington Library, San Marino, Calif.

13. Jaqueline Beidl, "The Blackfeet and the Badger–Two Medicine: An Evaluation of Potential Traditional Cultural Significance Drawn from Archival Sources," January 1992, unpublished paper on file in the George C. Ruhle Library, Glacier National Park, West Glacier, Mont. (hereafter Ruhle Library); Wissler, "Material Culture," 19–24; McClintock, *Old North Trail;* Beverly Hungry Wolf, *The Ways of My Grandmothers* (New York: William Morrow, 1980); Reeves and Peacock, "'Our Mountains Are Our Pillows,'" 1: 99–110. For a description of the buffalo skeletons, see Bailey, *Wild Animals,* 25–26.

14. Schultz, *Blackfeet Tales,* 98–109.

15. John C. Ewers, *The Blackfeet: Raiders on the Northwestern Plains* (Norman: University of Oklahoma Press, 1958), 277–312; William E. Farr, *The Reservation Blackfeet, 1882–1945: A Photographic History of Cultural Survival* (Seattle: University of Washington Press, 1984), 3–12 and passim; Michael F. Foley, "An Historical Analysis of the Administration of the Blackfeet Indian Reservation by the United States, 1855–1950's," Indian Claims Commission Docket No. 279-D [1974], 1–71; and Thomas R. Wessel, "Historical Report on the Blackfeet Reservation in Northern Montana," Indian Claims Commission Docket No. 279-D [1975], 1–47.

16. Chief White Calf's statement is recorded in George Bird Grinnell's journal entry for November 7, 1888; this and all subsequent references to Grinnell's journals come from "Journals," George Bird Grinnell Collection, Southwest Museum, Los Angeles.

17. Efforts to curb traditional practices could be quite severe and, as Blackfeet Agent L. W. Cooke described it, his job essentially boiled down to "Control [with] . . . a kind but unrelaxing guidance—a hand of steel in a glove of velvet": Cooke to the Commissioner of Indian Affairs (CIA), August 15, 1894, vol. 6, "Copies of Official Letters Sent, November 1878–June 1915," Records of the Blackfeet Agency, Bureau of Indian Affairs, record group 75, National Archives–Rocky Mountain Region, Denver.

18. Farr, *The Reservation Blackfeet*; Foley, "Historical Analysis," 72–93. For contemporary accounts of Blackfeet hunting in the Glacier area, see George Bird Grinnell's journal entries for November 7, 1888, and September 23–24, 1889; McClintock, *Old North Trail,* 14–15, 24, 47, and *The Tragedy of the Blackfoot* (Los Angeles: Southwest Museum, 1930), 8–13; and Warren L. Hanna, *The Life and Times of James Willard Schultz (Apikuni)* (Norman: University of Oklahoma Press, 1986), 138–144. Blackfeet elder Mike Swims Under recalls that his family depended upon the success of his father's hunting in the Rockies at the turn of the century; interview with the author, Heart Butte, Mont., August 8, 1994.

19. For a discussion of Grinnell's early career, see John F. Reiger, *American Sportsmen and the Origins of Conservation* (New York: Winchester Press, 1975), 132–151.

20. Quotations are from Grinnell's journal entry for September 11, 1891. Grinnell's articles about the Glacier region included a number of serials. See "The Rock Climbers," "Slide Rock from Many Mountains," and "Climbing Blackfoot," *Forest and Stream* 29 and 30 (December 29 to January 19 and January 26 to May 3, 1887–1888), 34 and 35 (March 6 and October 2, 1890), and 51 (October 8, 1898). Also see Gerald A. Diettert, *Grinnell's Glacier: George Bird Grinnell and Glacier National Park* (Missoula, Mont.: Mountain Press, 1992), 47–59.

21. Grinnell, journal entry for September 17, 1891. Shortly after returning to New York, Grinnell wrote "The Crown of the Continent," which he submitted to *Century Magazine.* While the article contained his first public appeal for the creation of Glacier National Park, it was not published until 1901. See Grinnell, "The Crown of the Continent," *Century Magazine* 62 (1901): 660–672; and Diettert, *Grinnell's Glacier,* 77.

22. For a discussion of Grinnell's efforts to "preserve" Indian culture and root out corruption in the Indian service, see Brian W. Dippie, *The Vanishing American: White Attitudes and U.S. Indian Policy* (Lawrence: University Press of Kansas, 1982), 222–228. Grinnell wrote often about the necessity of wilderness recreation for urban Americans and the importance of "civilization" programs for Indians. See, for example, Grinnell and Theodore Roosevelt, eds., *Trail and Camp-Fire* (New York: Forest and Stream Publishing, 1897). Also see Grinnell, "Opening Up Forest Reserves," *Forester* 4 (1898): 42–44; *Blackfoot Lodge Tales,* xiii–ix; and *The Story of the Indian* (New York: D. Appleton, 1895), ix–x.

23. Christopher S. Ashby, "The Blackfeet Agreement of 1895 and Glacier National Park: A Case History" (master's thesis, University of Montana, 1985), 20–21; Wessell, "Historical Report," 90–95; Foley, "Historical Analysis," 181–187; and Diettert, *Grinnell's Glacier,* 61–64.

24. Diettert, *Grinnell's Glacier,* 63.

25. Little Dog quoted in *Letter from the Secretary of the Interior, Transmitting an Agreement Made and Concluded Sept. 26, 1895, with the Indians of the Blackfeet Reservation,* 54th Cong., 1st sess., 1896, S. Doc. 118, 9.

26. Ibid, 12.

27. No record exists of these unofficial meetings, but some idea about what occurred can be gained from White Calf's comments in ibid., 18 and from Grinnell's journal entry for September 24, 1895.

28. For numbers of absent Blackfeet, see Grinnell's journal entry for September 28, 1895.

29. S. Doc. 118, 18.

30. Ibid, 19.

31. Phillip E. Roy, "Position Paper of the Blackfeet Tribe of the Blackfeet Indian Reservation: Regarding That Portion of the Lewis and Clark National Forest Which Was in 1896 Divested from the Blackfeet Indian Reservation by an Act of Congress," September 19, 1979, 8. I am grateful to Chief Earl Old Person, chairman of the Blackfeet Tribal Council, for providing me with a copy of this document.

32. S. Doc. 118, 21.

33. Ibid.

34. Louis S. Warren, *The Hunter's Game: Poachers and Conservationists in Twentieth-Century America* (New Haven: Yale University Press, 1997), 133. As Warren notes, the ban was later rescinded in 1897. For a discussion of the probable connection between the Jackson Hole case and the 1895 agreement with the Blackfeet, see Kenneth Pitt, "The Ceded Strip: Blackfeet Treaty Rights in the 1980s," unpublished paper in Ruhle Library.

35. Ted Hall, interview with the author, Browning, Mont., August 2, 1994; Roy, "Position Paper," 8–10. For evidence that the agreement may have been understood as a fifty-year lease, see the comments of Big Brave, who speaks of no more dealings with the government for fifty years, and White Calf, who intimates that he agreed to sell only the land above the timber line; S. Doc. 118, 19.

36. Senate ratification came on June 10, 1896. Consequently, the agreement between the commissioners and the Blackfeet is variously referred to as the "1895 Agreement" and the "1896 Agreement." In keeping with the actual dates of the meetings with the Blackfeet and to avoid confusion, I have chosen to use the earlier year in all subsequent references.

37. Diettert, *Grinnell's Glacier,* 73–75; and Sheire, *Glacier National Park,* 143–144. Muir joined the survey as an ex officio member.

38. President Grover Cleveland, "Proclamation to Establish the Lewis and Clark Forest Reserve in Northern Montana," *U.S. Statutes at Large,* 31 (1897): 911.

39. McClintock, *Old North Trail,* 20.

40. Diettert, *Grinnell's Glacier,* 80–95; Michael G. Schene, "The Crown of the Continent: Private Enterprise and Public Interest in the Early Development of Glacier National Park, 1910–1917," *Forest and Conservation History* 34 (1990): 69–75.

41. See, for example, Robert J. Hamilton to Louis W. Hill, March 19, 1912, file "Roads—Indians Support," Ruhle Library. Hamilton, an important tribal leader from 1912 to 1930, told Hill that the Blackfeet "highly appreciate [your] plans for the development . . . of the Glacier National Park" and assured him that the "Indians are [in] no way to obstruct your work, but are ready to support you at any old time."

42. Quotation is from *U.S. Statutes at Large* 277 (1910), 354.

Notes to Chapter 6

1. Tom Dillon, *Over the Trails of Glacier National Park* (St. Paul: Great Nothern Railway, [1912]), 5–6.

2. Quotes from various advertisements for the promotion of Glacier tourism in Great Northern Railway Company, Advertising and Publicity Department, "Magazine and Newspaper Advertisements, 1884–1970," microfilm edition, vol. 1, rolls 1 and 6, Minnesota Historical Society, Minneapolis. The long fascination with the so-called vanishing Indian and its effect on government policy is well documented in Brian W. Dippie, *The Vanishing American: White Attitudes and U.S. Indian Policy* (Lawrence: University Press of Kansas, 1982).

3. For an excellent study of Great Northern's promotion of Glacier National Park, see Marguerite S. Shaffer, "See America First: Tourism and National Identity, 1905–1930" (Ph.D. diss., Harvard University, 1994), 92–143. Also see Ann T. Walton, "The Louis W. Hill, Sr., Collection of American Indian Art," in *After the Buffalo Were Gone: The Louis Warren Hill Collection of Indian Art,* ed. Ann T. Walton, John C. Ewers, and Royal B. Hassrick (St. Paul: Northwest Area Foundation, 1985), 11–35; Ann Regan, "The Blackfeet, the Bureaucrats, and Glacier National Park, 1910–1940," paper presented at the Western History Association conference, Billings, Mont., October 1986, copy in the George C. Ruhle Library, Glacier National Park, West Glacier, Mont. (hereafter Ruhle Library).

4. Edward Frank Allen, "The Greatness of Glacier National Park," *Travel* 20 (1913): 9–11.

5. Ibid., 12–13; Robert Sterling Yard, *National Park Portfolio* (New York: Scribner's Sons, 1916). Also see Dillon, *Over the Trails,* 5–7.

6. John F. Sears, *Sacred Places: American Tourist Attractions in the Nineteenth Century* (New York: Oxford University Press, 1991). Also see Anne Farrar Hyde, *An American Vision: Far Western Landscape and National Culture, 1820–1920* (New York: New York University Press, 1990); and Jules Prown et al., *Discovered Lands, Invented Pasts: Transforming Visions of the American West* (New Haven: Yale University Press, 1992). For the larger context of nationalism and nationalistic symbols in the late nineteenth century, see Eric Hobsbawm, *Nations and Nationalism since 1780: Programme, Myth, Reality,* 2d ed. (Cambridge: Cambridge University Press, 1990); and Benedict Anderson, *Imagined Communities: Reflections on the Origin and Spread of Nationalism,* 2d ed. (London: Verso, 1991).

7. Quotation is from *U.S. Statutes at Large* 2777 (1910): 354.

8. Mary Roberts Rinehart exactly expressed these sentiments in her book *Tenting To-Night: A Chronicle of Sport and Adventure in Glacier Park and the Cascade Mountains* (Boston: Houghton Mifflin, 1918), 96–99.

9. Though reprinted in countless promotional brochures, this quote originally comes from Rinehart, *Through Glacier Park: Seeing America First with Howard Eaton* (Boston: Houghton Mifflin, 1916), 24–25.

10. Roderick Nash, *Wilderness and the American Mind,* 3d ed. (New Haven: Yale University Press, 1982), 141–160. The term "intense experience" comes from T. J. Jackson Lears, *No Place of Grace: Antimodernism and the Transformation of American Culture, 1880–1920* (New York: Pantheon, 1981); also see Tom Lutz, *American Nervousness, 1903: An Anecdotal History* (Ithaca, N.Y.: Cornell University Press, 1991).

11. John Muir coined *overcivilized* in *Our National Parks* (Boston: Houghton Mifflin, 1901), 1.

12. Theodore Roosevelt, *The Wilderness Hunter* (New York: G. P. Putnam's Sons, 1893), xxxi; Rinehart, *Through Glacier Park,* 23–24; Ada F. Chalmers, "Through Glacier Park for $1.39 a Day: How I Did It with My Two Boys," *The Ladies Home Journal* 34 (1917): 65–66; Lulie Nettleton, *With the "Mountaineers" in Glacier National Park: Walking Tours Book* (St. Paul: Great Northern Railway Company, 1915), copy in the William Andrews Clark Memorial Library, Los Angeles. The Great Northern went to tremendous lengths to appeal to the "New Woman," producing a vast array of brochures, magazine advertisements, and pamphlets that featured young women fishing, hiking, and climbing throughout the park. For a broader examination of the New Woman, see Rosalind Rosenberg, *Beyond Separate Spheres: Intellectual Roots of Modern Feminism* (New Haven: Yale University Press, 1982), 54–83. Also see Polly Welts Kaufman, *National Parks and the Woman's Voice: A History* (Albuquerque: University of New Mexico Press, 1996), 21, 69.

13. *U.S. Statutes at Large* 2777 (1910): 354. Some of the many scientific publications produced by the government and sold to tourists included Marius R. Campbell, *Origin of the Scenic Features of the Glacier National Park* (Washington, D.C.: Government Printing Office, 1914); Campbell, *The Glacier National Park: A Popular Guide to Its Geology and Scenery* (Washington D.C.: General Printing Office, 1914); and Vernon Bailey, *Wild Animals of Glacier National Park* (Washington D.C.: Government Printing Office, 1918). Game preservation was a key factor in winning support for the creation of Glacier and other national parks; see John F. Reiger, *American Sportsmen and the Origins of Conservation* (New York: Winchester Press, 1975), 93–113.

14. Richard West Sellars provides an excellent study of predator reduction programs at Glacier and other national parks; see his *Preserving Nature in the National Parks: A History* (New Haven: Yale University Press, 1997), 69–82.

15. Eagles were killed to increase the number of mountain goats in the park. For rangers' estimates on game numbers and the progress of predator reduction programs in the early 1910s, see George Bird Grinnell's entries for July 12, 1911; August 10 and 12, 1912; and August 8, 1915, in "Journals," George Bird Grinnell Collection, Southwest Museum, Los Angeles.

16. Henry W. Hutchings to Secretary of the Interior, April 4, 1912, box 23, folder "Game Protection;" central files 1907–1939; General Records of the National Park Service, Record Group 79; National Archives, Washington, D.C. (hereafter NA RG 79, NPS, "Game Protection"). I am indebted to Louis Warren for bringing this source to my attention and for providing me with a copy of his dissertation. The following discussion of National Park Service efforts to halt Blackfeet hunting on park lands parallels his study in several particulars. See Louis S. Warren, *The Hunter's Game: Poachers and Conservationists in Twentieth-Century America* (New Haven: Yale University Press, 1997), 126–171.

17. Efforts to improve the viewing of game populations through the planting of hay and the strategic placement of salt licks are discussed in Warren, *Hunter's Game,* 141. For an overview of poaching on the western portions of Glacier National Park, see ibid., 136–140.

18. Stephen T. Mather to solicitor for the Interior Department, November 4, 1915, NA RG 79, NPS, "Game Protection."

19. "Report of Game Killed and Trespass committed in Glacier National Park, October 1st to November 30th, Inclusive, Year 1913," NA RG 79, NPS, "Game Protection."

20. On private inholdings, see Sellars, *Preserving Nature,* 65–66; and Warren, *Hunter's Game,* 139–140.

21. William R. Logan, *Report of the Superintendent of the Glacier National Park* (Washington, D.C.: Government Printing Office, 1911); R. H. Chapman, *Report of the Superintendent of the Glacier National Park* (Washington, D.C.: Government Printing Office, 1912).

22. Acting superintendent to Glacier National park rangers, September 24, 1912, box 15, folder 95, "Superintendent's Subject Files, 1908–1929," Records of the Blackfeet Agency, Bureau of Indian Affairs, record group 75, National Archives–Rocky Mountain Region, Denver (hereafter NA-RMR RG 75, BIA SSF, 1908–1929, Blackfeet). In their encounters with rangers, the Blackfeet often claimed ignorance about park rules and boundaries; see Clemet S. Ucker to Chapman, August 1, 1912, NA RG 79, NPS, "Game Protection."

23. James L. Galen, *Report of the Superintendent of the Glacier National Park to the Secretary of the Interior* (Washington, D.C.: Government Printing Office, 1913), 15; also see Ucker to Chapman, June 21, 1912, NA RG 79, NPS, "Game Protection."

24. E. B. Meritt to C. L. Lewis, August 3, 1915, box 13, folder 88, NA-RMR RG 75, BIA SSF, 1908–1929, Blackfeet.

25. Lewis to commissioner of Indian Affairs (CIA), July 18, 1915, box 11, folder 26; Lewis to the Indians of the Blackfeet Tribe, January 26, 1916, box 13, folder 88; Lewis to Ralston, January 27, 1916, box 13, folder 123; Lewis to CIA, September 29, 1916, box 11, folder 25; all in NA-RMR RG 75, BIA SSF, 1908–1929, Blackfeet.

26. Quote from Arthur E. McFatridge to CIA, March 2, 1914, file 68193, Blackfeet 013, "Central Classified Files, 1907–1939," Records of the Bureau of Indian Affairs, Record Group 75, National Archives, Washington, D.C. (hereafter NA RG 75, BIA CCF, 1907–39, Blackfeet). Also see Blackfeet Tribal Delegation to Cato Sells, March 17, 1914, NA RG 75, BIA CCF, 1907–39, Blackfeet 013/68193; Lewis to Ralston, January 27, 1916, NA RG 75, BIA CCF, 1907–39, Blackfeet 013/123; and Lewis to CIA, September 29, 1916, box 11, folder 25, NA-RMR RG 75, BIA SSF, 1908–1929, Blackfeet.

27. Some 150 to 200 Blackfeet and their families hunted in the park area, and their efforts contributed to the liveihood of between 750 to 1,000 Indians, of an entire reservation population of approximately 2,300. See *Court of Claims of the United States, No. E-427: Blackfeet* (et al.) *Indians v. The United States, Evidence for Plaintiffs* (Washington D.C.: Government Printing Office, 1929), 119–127; and Christopher S. Ashby, "The Blackfeet Agreement of 1895 and Glacier National Park: A Case History" (master's thesis, University of Montana, 1985), 49–50.

28. D. D. LaBreche to Harry Lane, November 1, 1915, NA RG 75, BIA CCF, 1907–39, Blackfeet 115/119292.

29. Peter Oscar Little Chief to Harry Lane, November 20, 1915, NA RG 79, NPS, "Game Protection."

30. I. McBride to Meritt, November 6, 1915; Bo Sweeney to CIA, January 6, 1916; both in NA RG 75, BIA CCF, 1907–39, Blackfeet 115/119292. Also Mather to Solicitor, November 4, 1915; E. J. Ayers to Little Chief, November 27, 1915; and Ayers to CIA, November 27, 1915, all in NA RG 79, NPS, "Game Protection." While Mather was certainly aware of LaBreche, he had already expressed a concern about the legality of excluding Indians from the park when Glacier Superintendent Samuel F. Ralston brought the matter to his attention that previous summer.

31. Preston C. West to secretary of the Interior, January 4, 1916; Meritt to C. L. Ellis,

January 14, 1916, both in NA RG 75, BIA CCF, 1907–39, Blackfeet 115/119292. Also, Sweeney to Ralston, January 6, 1916, NA RG 79, NPS, "Game Protection."

32. West to Secretary of the Interior, January 4, 1916. For a discussion of the decision in *Ward v. Race Horse* (1896), see chapter 4.

33. CIA to Thomas Ferris, Supt., July 30, 1917; Acting Director of the National Park Service (NPS) to CIA, July 2, 1917; Ferris to CIA, November 19, 1917; Meritt to Mather, December 3, 1917; CIA to Director, NPS, January 29, 1918; and CIA to Ferris, August 22, 1917, all in NA RG 75, BIA CCF, 1907–39, Blackfeet 115/66598. Also, Lane to Senator H. L. Meyers, January 19, 1918, File Glacier National Park 12-14, box 2005, folder "Wild Animals," Central Classified Files 1907–1936, Office of the Secretary of the Interior, Record Group 48, National Archives, Washington, D.C. (hereafter NA RG 48, CCF, 1907–36, "Wild Animals").

34. A subsequent report from the Indian Field Service showed that some 338 Blackfeet owned land in the area between the park boundary and the highway. See O. A. Waetjen to F. L. Carter, Chief Ranger, Glacier National Park, April 11, 1928, Glacier National Park Archives, West Glacier, Mont. (hereafter GNPA). In the summer of 1994, when I conducted my research in Glacier National Park, the park archives were in the process of being recatalogued. Consequently, I have not been able to cite records according to file or box numbers.

35. Merrit to Ferris, November 5, 1917, NA RG 75, BIA CCF, 1907–39, Blackfeet 115/66598. Also, Acting director, NPS to CIA, July 2, 1917; and Ferris to CIA, November 19, 1917, both in NA RG 75, BIA CCF, 1907–39, Blackfeet 115/66598.

36. George E. Goodwin, "Glacier National Park," in *Report of the Director of the National Park Service to the Secretary of the Interior* (Washington, D.C.: Government Printing Office, 1917), 182–184.

37. CIA to Ferris, July 30, 1917, NA RG 75, BIA CCF, 1907–39, Blackfeet 115/66548. Also Merrit to Stephen Mather, December 3, 1917; and CIA to Director, NPS, January 29, 1918, both in NA RG 75, BIA CCF, 1907–39, Blackfeet 115/66598.

38. William T. Hornaday to Lane, April 7, 1919; George B. Grinnell to Hornaday, April 7, 1919; Horace Albright to Mr. Cotter, April 12, 1919, all in NA RG 75, BIA CCF, 1907–39, Blackfeet 115/66548.

39. Hornaday to Lane, April 9, 1919, and Lane to Hornaday, April 16, 1919, both in NA RG 48, CCF, 1907–36, "Wild Animals."

40. Merrit to Horace G. Wilson, April 22, 1919; Wilson to CIA, June 10, 1919, both in NA RG 75, BIA CCF, 1907–39, Blackfeet 115/66548.

41. Quoted in CIA, "Memorandum: Attention Mr. Cotter," January 15, 1918, NA RG 48, CCF, 1907–36, "Wild Animals."

42. For the remarks of Tail Feathers Coming Over the Hill, see James Willard Schultz, *Blackfeet Tales of Glacier National Park* (Boston: Houghton Mifflin, 1916), 1–2.

43. Chas. H. Burke to Fred C. Campbell, March 4, 1923, and enclosures, NA RG 75, BIA CCF, 1907–39, Blackfeet 115/17275, J. Ross Eakin, "Superintendent's Monthly Reports to the Director, National Park Service," January 10 and December 9, 1922, January 9 and July 10, 1923, GNPA.

44. Little Chief to Walsh, December 7, 1926; Little Chief to Walsh, January 23, 1928, with copy of 1924 petition attached, NA RG 75, BIA CCF, 1907–39, Blackfeet 115/7807. Also see Regan, "The Blackfeet," 10–11.

45. CIA to Walsh, March 21, 1928, NA RG 75, BIA CCF, 1907–39, Blackfeet 115/7807. Also se Warren, *Hunter's Game,* 318–322. The Indian Citizenship Act of 1924 conferred citizenship on all native peoples.

46. Ranger Station Log Books, "Daily Reports" for the Two Medicine Ranger Station, November and December 1931, GNPA.

47. E. T. Scoyen to Arthur P. Archer, December 31, 1932, box 238, file 7; Central Files 1907–1939; General Records of the National Park Service, Record Group 79; National Archives, Washigton, D.C.

48. "Petition of the Blackfeet Indians of Montana for a Correct Survey of the Western Boundary of the Blackfeet Reservation Between the Western End of the Lower Two Medicine Lake and Heart Butte," July 21, 1931, NA RG 75, BIA CCF, 1907–39, Blackfeet 307.2/1566.

49. Scoyen to Director, April 22, 1932, GNPA.

50. Scoyen to Mather, December 31, 1932; John H. Edwards to Attorney General, January 10, 1933; and Scoyen to Acher, December 31, 1932, all in box 238, file 7; Central Files 1907–1939; General Records of the National Park Service, Record Group 79; National Archives, Washington, D.C.

51. Little Dog to CIA, December 4, 1909, vol. 22, "Copies of Official Letters Sent, November 1878–June 1915," Records of the Blackfeet Agency, Bureau of Indian Affairs, Record Group 75, National Archives–Rocky Mountain Region, Denver; *An Act for the Relief of Certain Nations or Tribes of Indians in Montana, Idaho, and Washington, U.S. Statutes at Large* 42 (1924): 21; and *Court of Claims of the United States, No. E-427: Evidence,* 27–127.

52. E. C. Finney, "Solicitor's Opinion of the Blackfeet Rights on Glacier Park Land, 1932," June 21, 1932, Ruhle Library.

53. Scoyen to Acher, December 31, 1932.

54. *Blackfeet* (et al) *v. United States: Final Decision and Opinion* (Washington, D.C.: Government Printing Office, 1935), 10–11.

55. Stephen Mather to Cato Sells, December 13, 1919, NA RG 75, BIA CCF, 1907–39, Blackfeet 307.4/106653; Mather to Sells, January 7, 1921, GNPA.

56. Quote is from J. Ross Eakin, "Annual Report of the Superintendent of Glacier National Park, 1928," 14, Ruhle Library.

57. Albright to Ralph Budd, October 16 and 27, 1926; Budd to Albright, October 20, 1926; and Albright to Budd, February 15, 1927, all in GNPA. In conjunction with his efforts to include the Great Northern Railroad as an ally in the proposed park extension, Albright also corresponded with George Bird Grinnell, Representative Ralph Cramton, and Assistant Secretary of the Interior A. E. Demaray.

58. Albright to CIA, January 13, 1930; CIA to Albright, January 24, 1930, both in GNPA. For notes on earlier Blackfeet protests, see Burke to Senator Wheeler, December 17, 1928; and Cambell to CIA, January 4, 1929, both in NA RG 75, BIA CCF, 1907–39, Blackfeet 013/68193.

59. J. R. Eakin to the Director, NPS, March 24, 1930, GNPA.

60. Forrest R. Stone to Joseph M. Dixon, September 12, 1930, NA RG 75, BIA CCF, 1907–39, Blackfeet 307.2/50076.

61. Albright, "Memorandum to Superintendent Glacier National Park, February 16, 1931," NA RG 75, BIA CCF, 1907–39, Blackfeet 307.2/10501.

62. Albright to CIA, December 9, 1931, NA RG 75, BIA CCF, 1907–39, Blackfeet 013/68193.

63. C. J. Rhoads to Albright, January 2, 1932, NA RG 75, BIA CCF, 1907–39, Blackfeet 013/68193.

64. Stone to CIA, January 11, 1932, NA RG 75, BIA CCF, 1907–39, Blackfeet 013/68193.

65. Hal Rothman, *Preserving Different Pasts: The American National Monuments* (Urbana: University of Illinois Press, 1989), 141.

66. Scoyen to Director, NPS, April 23, 1935; Scoyen to Director, NPS, June 29, 1935, both in GNPA. The amount of correspondence generated by this subject over the years is enormous. The best summary of efforts to expand Glacier National Park eastward is

"Glacier National Park, Montana: A History of Its Establishment and Revision of its Boundaries," n. d., 6–19, Ruhle Library.

67. For Blackfeet adoption of the Indian New Deal, see John C. Ewers, *The Blackfeet, Raiders on the Northwestern Plains* (Norman: University of Oklahoma Press, 1958), 323–324. The standard work on the Indian New Deal remains Graham D. Taylor, *The New Deal and American Indian Tribalism: The Administration of the Indian Reorganization Act, 1934–45* (Lincoln: University of Nebraska Press, 1980).

68. Scoyen, "Annual Report of the Superintendent of Glacier National Park, 1937," 15–17, Ruhle Library.

69. J. W. Emmert, Memorandum for Regional Director, September 22, 1944, GNPA.

70. Emmert, Memorandum for Regional Director, April 3, 1953, GNPA. For a thorough study of game management in Glacier National Park and changing attitudes about Blackfeet hunters in the 1940s and 1950s, see Warren, *Hunter's Game,* 163–165.

71. Ted Hall, interview with the author, Browning, Mont., July 14, 1994. The comments of Big Brave at the 1895 negotiations suggest an understanding of a fifty-year lease. Chief White Calf also intimated that he agreed to sell only the land above the timber line. Both quoted in *Letter from the Secretary of the Interior, Transmitting an Agreement Made and Concluded Sept. 26, 1895, with the Indians of the Blackfeet Reservation,* 54th Cong., 1st Sess., 1895, Doc. 118, 19.

72. Ashby, "The Blackfeet Agreement of 1895," 57–60.

73. For a contemporary environmentalist critique of Blackfeet actions, see "Triple Jeopardy at Glacier National Park" and "Glacier: Beleaguered Park of 1975," *The National Parks & Conservation Magazine,* September 1975, 20–22 and November 1975, 4–10.

74. Retired ranger (anonymous), interview with the author, Santa Barbara, Calif., November 23, 1994.

75. "Blackfeet Object to Proposed Park Fence," *Columbia Falls Tribune,* 19 August, 1978, p. 1.

76. For an overview of relations between the Blackfeet and the National Park Service in the 1980s, see Kenneth P. Pitt, "The Ceded Strip: Blackfeet Treaty Rights in the 1980s," unpublished paper in Ruhle Library.

77. Albright to CIA, January 13, 1930.

Notes to Chapter 7

1. Frederick Law Olmsted, "The Yosemite Valley and the Mariposa Big Trees: A Preliminary Report," *Landscape Architecture* 43 (1952): 22.

2. Lafayette H. Bunnell, *Discovery of the Yosemite, and the Indian War of 1851, Which Led to That Event* (Chicago: Fleming H. Revel, 1880); Carl P. Russell, *One Hundred Years in Yosemite: The Story of a Great Park and Its Friends* (Yosemite National Park: Yosemite Natural History Association, 1957), 36–48; Alfred Runte, *Yosemite: The Embattled Wilderness* (Lincoln: University of Nebraska Press, 1990), 10–12; Craig D. Bates and Martha Lee, *Tradition and Innovation: A Basket History of the Indians of the Yosemite-Mono Lake Area* (Yosemite National Park: Yosemite Association, 1991), 26–27. Two works that associate efforts to remove the Yosemite Indians from the area in the 1850s with the later development of Yosemite National Park are Susanna Hecht and Alexander Cockburn, *The Fate of the Forest: Developers, Destroyers, and Defenders of the Amazon* (New York: Harper Collins, 1990), 269–276; and Rebecca Solnit, *Savage Dreams: A Journey into the Hidden Wars of the American West* (San Francisco: Sierra Club Books, 1994).

3. Russell, *One Hundred Years in Yosemite,* 40; Jean-Nicholas Perlot, *Gold Seeker: Adventures of a Belgian Argonaut During the Gold Rush Years,* trans. Helen Harding Bretnor, ed. Howard R. Lamar (New Haven: Yale University Press, 1985), 228. For a general account of

California Indians working in the gold country, see James J. Rawls, "Gold Diggers: Indian Miners in the California Gold Rush," *California Historical Quarterly* 55 (1976): 28–45; and Albert Hurtado, *Indian Survival on the California Frontier* (New Haven: Yale University Press, 1988), 100-117.

4. James M. Hutchings, *In the Heart of the Sierras: The Yo-Semite Valley* (Oakland: Pacific Press, 1886), 100.

5. Perlot, *Gold Seeker,* 294.

6. H. Willis Baxley, *What I Saw on the West Coast of South and North America and the Hawaiian Islands* (New York: D. Appleton, 1865), 467.

7. Hutchings, *In the Heart of the Sierras,* 130.

8. Runte, *Yosemite,* 51-54.

9. Hutchings, *In the Heart of the Sierras,* 130; Bates and Lee, *Tradition and Innovation,* 31.

10. Karen P. Wells and Craig D. Bates, "Ethnohistory and Material Culture of Southern Sierra Miwok: 1852–1880," unpublished essay in the Yosemite National Park Research Library (hereafter YNPRL); C. E. Kelsey, *Census of Non-Reservation California Indians, 1905-1906,* ed. Robert Heizer (Berkeley: University of California Archaeological Research Facility, Department of Anthropology, 1971); James Gary Maniery, *Six Mile and Murphy's Rancherias: An Ethnohistorical and Archaeological Study of Two Central Sierra Miwok Village Sites* (San Diego: San Diego Museum of Man, 1987); Eugene L. Conrotto, *Miwok Means People: The Life and Fate of the Native Inhabitants of the Gold Rush Country* (Fresno: Valley Publishers, 1973), 97–98. The Washo Indians of the Lake Tahoe area made similar adaptations to agricultural and mining developments in their homeland; see James F. Downs, *Two Worlds of the Washo: An Indian Tribe of California and Nevada* (New York: Holt, Rinehart and Winston, 1966).

11. Bates and Lee, *Tradition and Innovation,* 15–19. Quote is from Galen Clark, *Indians of the Yosemite Valley and Vicinity: Their History, Customs, and Traditions, with an Appendix of Useful Information for Yosemite Visitors* (Yosemite Valley, Calif.: Galen Clark, 1904), 80. Clark derived his source from fragments of stories told over fifty years. Also see Samuel A. Barrett, "Myths of the Southern Sierra Miwok," *University of California Publications in American Archaeology and Ethnology* 16 (1919), 1–28; C. Hart Merriam, *Dawn of the World: Weird Tales of the Mewan Indians of California* (Cleveland: Arthur H. Clark, 1910; reprinted as *The Dawn of the World: Myths and Tales of the Miwok Indians of California,* (Lincoln: University of Nebraska Press, 1993), 93, 229; Stephen Powers, *Tribes of California* (Washington, D.C.: Government Printing Office, 1877; reprint, Berkeley: University of California Press, 1976), 362–367; and Samuel A. Barrett and Edward W. Gifford, "Miwok Material Culture," in *Bulletin of the Public Museum of the City of Milwaukee* (Milwaukee: Public Museum of the City of Milwaukee, 1933; reprint, Yosemite National Park: Yosemite Association, 1990).

12. Bates and Lee, *Tradition and Innovation,* 19.

13. Ibid., 23, 25–26.

14. Lafayette H. Bunnell, from a letter quoted in Hutchings, "The Yo-Ham-i-te Valley," *Hutchings' California Magazine* 1 (1856): 7.

15. Cultural Systems Research, Inc., "Petition to the Government of the United States from the American Indian Council of Mariposa County for Acknowledgment as the Yosemite Indian Tribe" (draft), prepared by Lowell John Bean and Sylvia Brakke Vane (Menlo Park, Calif.: Cultural Systems Research, Inc., 1984), copy in YNPRL (hereafter CSRI, "Petition to the Government"); Wells and Bates, "Ethnohistory and Material Culture." Efforts on the part of park officials and scholars to define "true" Yosemite Indians as directly descended from the Ahwahneechee have always been flawed because such a definition implies that a static culture once existed and then perished when change arrived in the midnineteenth century. In much the same way that the Puritans of the seventeenth century are gone, one could claim that the Ahwahneechee of the early 1800s no longer exist. But just as there have been New Englanders for nearly five centuries, so, too, have

there been Yosemite Indians since the last Ice Age. Though much different from even two generations ago, the Yosemite Indians remain a distinct and dynamic cultural group today, with close ties to Yosemite Valley and the surrounding area.

16. See J. F. Campbell, *My Circular Notes. Extracts from Journals, Letters Sent Home, Geological and Other Notes Written While Traveling Westwards round the World from July 6, 1874–July 6, 1875* (London: Macmillan, 1876), 79; Charles Carleton Coffin, *Our New Way Round the World* (Boston: Fields, Osgood, 1869), 478; Samuel Kneeland, *The Wonders of the Yosemite Valley, and of California* (Boston: Alexander Moore, Lee, and Shepard, 1872), 52–53; Rev. W. W. Ross, *10,000 Miles by Land and Sea* (Toronto: James Campbell, 1876), 180; A. E. Wood, *Annual Report of the Acting Superintendent of the Yosemite National Park to the Secretary of the Interior* (Washington, D.C.: Government Printing Office, 1892), 14–15; Charles Francis Saunders, *Under the Sky in California* (New York: McBride, Nast, 1913), 69.

17. John S. Hittell, *Yosemite: Its Wonders and Its Beauties* (San Francisco: H. H. Bancroft, 1868), 30.

18. Bates and Lee, *Tradition and Innovation,* 34.

19. Saunders, *Under the Sky in California,* 64.

20. See Robert F. Heizer, *The Destruction of the California Indians* (Santa Barbara: Peregrine Smith, 1974), 11; Hurtado, *Indian Survival on the California Frontier,* 100–117, 149–168. It is generally agreed that California's Indian population fell from 150,000 in 1848 to 35,000 in 1860, declining to a nadir of some 20,000 by the 1890s. Sierra Miwok numbers plummeted even more drastically, from an estimated 9,000 in aboriginal times to 760 individuals in 1910. This profound demographic collapse resulted primarily from starvation, disease, and murder. See Sherburne F. Cook, *The Population of the California Indians, 1769–1970* (Berkeley: University of California Press, 1976), 43–73; Alfred Kroeber, "Indians of Yosemite," in *Handbook of Yosemite National Park,* ed. A. F. Hall (New York: G. P. Putnam and Sons, 1921), 54; Russell Thornton, *American Indian Holocaust and Survival: A Population History since 1492* (Norman: University of Oklahoma Press, 1987), 107–113.

21. Choko was also known as Old Jim; quoted in Powers, *Tribes of California,* 368.

22. Wells and Bates, "Ethnohistory and Material Culture." For studies of federal and state Indian policies in California, see Joseph Ellison, *California and the Nation, 1850-1869: A Study of the Relations of a Frontier Community with the Federal Government* (Berkeley: University of California Press, 1927), 97–102; Hurtado, *Indian Survival on the California Frontier,* 126–148; Rawls, *Indians of California: The Changing Image* (Norman: University of Oklahoma Press, 1984) 137–160; and George Harwood Phillips, *Indians and Indian Agents: The Origins of the Reservation System in California, 1849–1852* (Norman: University of Oklahoma Press, 1997).

23. Kneeland, *Wonders of the Yosemite Valley,* 52.

24. Although a number of tourist accounts mention the Indian settlement at Wawona, see especially John Erastus Lester, *The Atlantic to the Pacific: What to See and How to See It* (Boston: Shepard and Gill, 1873), 140; and Anonymous, *Souvenir of Yosemite* (n.p., 1886?), 7.

25. Charles B. Turrill, *California Notes* (San Francisco: Edward Bosqui, 1876), 223–224; Campbell, *My Circular Notes,* 79; Lester, *Atlantic to the Pacific,* 140, 156; Ross, *10,000 Miles by Land and Sea,* 181–182; "Digger Indian Fare," *San Francisco Chronicle,* 12 July 1889, p. 4; Saunders, *Under the Sky in California,* 67.

26. *Mariposa Gazette,* 11 October 1855, quoted in Peter Browning, *Yosemite Place Names: The Historic Background of Geographic Names in Yosemite National Park* (Lafayette, Calif.: Great West Books, 1988), 216.

27. George W. Pine, *Beyond the West* (Utica, N.Y.: T. J. Griffiths, 1871), 417.

28. Kate Nearpass Ogden, "Sublime Vistas and Scenic Backdrops: Nineteenth-Century Painters and Photographers at Yosemite," *California History* 69 (1990): 146–163.

29. C[onstance] F[letcher] Gordon Cumming, *Granite Crags of California* (Edinburgh: William Blackwood and Sons, 1886), 137.

30. Hutchings, *In the Heart of the Sierras,* 421–422.

31. Bates and Lee, *Tradition and Innovation,* 73–89.

32. Ibid., 8. For a contemporary comment on this phenomenon, see Clark, *Indians of the Yosemite Valley,* 1.

33. Bates and Lee, *Tradition and Innovation,* 22–23.

34. Because of their marked importance to tourists and the members of their own community, these women or their families often served as intermediaries between park officials and the Yosemite Indians. Important early basket makers were often identified as Indian leaders, and their status within the Yosemite Indian community often passed to their daughters, sons, and grandchildren. Ibid., 143.

35. Although early tourists commented on such "fandangos," it is difficult to determine whether they were part of a ceremony that outsiders were invited to witness—perhaps for a fee—or if they were simply a commercial event. For a comment on individual dancers and singers, see Helen Hunt Jackson, *Bits of Travel at Home and Abroad* (Boston: J. R. Osgood, 1894), 107.

36. Anonymous, *A Souvenir of Yosemite,* 7.

37. Frank T. Lea, "Indian Bread Makers in Yosemite," *Overland Monthly* 64 (1914): 25.

38. Clark, *Indians of the Yosemite Valley,* 104.

39. J. Smeaton Chase, *Yosemite Trails: Camp and Pack-Train in the Yosemite Region of the Sierra Nevada* (Boston: Houghton Mifflin, 1911), 32–33.

40. Kneeland, *The Wonders of the Yosemite Valley,* 52.

41. Moses Harris, *Report of the Superintendent of the Yellowstone National Park, 1889* (Washington, D.C.: Government Printing Office, 1889), 13–16. See chapter 4.

42. Wood, *Annual Report,* 13.

43. Thomas Starr King, *A Vacation Among the Sierras: Yosemite in 1860,* ed. John A. Hussey (San Francisco: Book Club of California, 1962), 40–41. See also Stanford E. Demars, *The Tourist in Yosemite, 1855–1985* (Salt Lake City: University of Utah Press, 1991), 38–39. The gathering of acorns from woodpecker stores generally occurred only in times of extreme shortage; Barrett and Gifford, "Miwok Material Culture," 143. I am especially grateful to Dave Raymond for pointing out a previous error in my analysis of Starr King and his observations of Miwok behavior.

44. Jackson, *Bits of Travel,* 107.

45. John Muir, *The Mountains of California* (New York: Century, 1894), 93.

46. Edward Castillo, "Petition to Congress on Behalf of the Yosemite Indians," *Journal of California Anthropology* 5 (1978): 271–272.

47. Craig Bates and Martha Lee (*Tradition and Innovation,* 36) draw a similar conclusion about Robinson's efforts on behalf of the Yosemite Indians.

48. California Legislature, Assembly Committee on Yosemite Valley and Mariposa Big Trees, *In the Matter of the Investigation of the Yosemite Valley Commissioners,* 28 sess., February 1889 (Sacramento: State Printing Office, 1889); Runte, *Yosemite,* 57.

49. Runte, *Yosemite,* 61–63.

50. For evidence of Yosemite hunting in these areas, see Coffin, *Our New Way Round the World,* 478; Muir, *The Mountains of California,* 80–81; and Muir, *My First Summer in the Sierra* (Boston: Houghton Mifflin, 1916), 94.

51. Alexander Rodgers, *Annual Report of the Acting Superintendent of the Yosemite National Park to the Secretary of the Interior* (Washington D.C.: Government Printing Office, 1898).

52. Saunders, *Under the Sky in California,* 69–70. For very similar observations and conclusions, see "She Was Simply a Yosemite Squaw," *Yosemite Tourist,* 25 June 1907, p. 3.

53. Quote is from David Sherfey to William Littebrandt, December 5, 1913; enclosed in a letter from Littebrandt to Secretary of the Interior, July 11, 1914, file 78250, Yosemite 050, "Central Classified Files, 1907–1939," Records of the Bureau of Indian Affairs,

Record Group 75, National Archives, Washington, D.C. (hereafter NA RG 75, BIA CCF, 1907–39, Yosemite).

54. C. H. Asbury to Commissioner of Indian Affairs (CIA), September 22, 1914, NA RG 75, BIA CCF, 1907–39, Yosemite 050/78250.

55. "Memorandum for Chief Education Division, Yosemite," January 29, 1915, NA RG 75, BIA CCF, 1907–39, Yosemite 050/78250.

56. Reference to "old Indian camp" is from B. B. Marshall to Stephen T. Mather, October 13, 1915, box 308, folder "Yosemite National Park: Repairs and Improvements, Yosemite Village"; Central Files 1907–1939; General Records of the National Park Service, Record Group 79; National Archives, Washington, D.C.

57. Littebrandt to Secretary of the Interior, July 11, 1914.

58. Asbury to CIA, September 22, 1914.

59. The Yosemite Indians still do not have federally recognized tribal status, a situation that the American Indian Council of Mariposa County has sought to change for nearly two decades. For a brief overview of these current efforts, see Kat Anderson, "We Are Still Here," *Yosemite* 53 (1991), 1–5.

60. Asbury to CIA, January 22, 1915, NA RG 75, BIA CCF, 1907–39, Yosemite 050/78250.

Notes to Chapter 8

1. Mrs. H. J. [Rose Schuster] Taylor, *The Last Survivor* (San Francisco: Johnck & Seeger, 1932), 5, [ii].

2. On the Hetch Hetchy controversy, see Kendrick A. Clements, "Politics and the Park: San Francisco's Fight for Hetch Hetchy, 1908–1913," *Pacific Historical Review* 48 (1979): 185–215; Norris Hundley Jr., *The Great Thirst: Californians and Water, 1770s–1990s* (Berkeley: University of California Press, 1992), 170–185; Michael L. Smith, *Pacific Visions: California Scientists and the Environment, 1850–1915* (New Haven: Yale University Press, 1987), 172–185; and Alfred Runte, *Yosemite: The Embattled Wilderness* (Lincoln: University of Nebraska Press, 1990), 80–82.

3. Quotations of the National Park Service Act of 1916 are from Lary M. Dilsaver, ed., *America's National Park System: The Critical Documents* (Lanham, Md.: Rowman & Littlefield, 1994), 46–47.

4. Franklin K. Lane to Stephen T. Mather, May 18, 1917; quoted in Richard West Sellars, *Preserving Nature in the National Parks: A History* (New Haven: Yale University Press, 1997), 57.

5. Runte, *National Parks: The American Experience,* 3d ed. (Lincoln: University of Nebraska Press, 1997), 82–105; Dilsaver, ed., *America's National Park System,* 46–47. For an excellent discussion of predator control and its link to tourist expectations, see Sellars, *Preserving Nature,* 71–82.

6. Sellars, *Preserving Nature,* 47–90; Runte, *Yosemite,* 135–159.

7. Quotes are from W. B. Lewis to Stephen Mather, September 25, 1923, file "Indian Affairs" W 34, box 970.33: I-3, Yosemite National Park Research Library (hereafter "Indian Affairs," I-3, YNPRL).

8. For a broader treatment of American stereotypes of Native Americans, see Robert J. Berkhofer Jr., *The White Man's Indian: Images of the American Indian from Columbus to the Present* (New York, 1978); Brian Dippie, *The Vanishing Indian: White Attitudes and U.S. Indian Policy* (Lawrence: University Press of Kansas, 1982); Duane Allen Matz, "Images of Indians in American Popular Culture since 1865" (Ph.D. diss., Illinois State University, 1988); and Philip J. Deloria, *Playing Indian* (New Haven: Yale University Press, 1998).

9. Don Tressider to W. B. Lewis, June 18, 1925, file 883-07.3, "Indians-General, 1924 to 1931," box 970.33: I-5, (hereafter "Indians General," box subset and YNPRL).

10. Advertisement and prize list for Indian Field Days, 1925, "Indians-General" I-1, YNPRL.

11, Craig D. Bates and Martha Lee, *Tradition and Innovation: A Basket History of the Indians of the Yosemite–Mono Lake Area* (Yosemite National Park: Yosemite Association, 1991), 104–107. Because the 1929 Field Days lost money and the stock market crashed that autumn, and because the "commercial" nature of the event drew considerable criticism in the late 1920s, the field days were eliminated.

12. Handwritten letter addressed to "Chief Townsley," August 5, 1924, "Indians General " I-1, YNPRL. Subsequent correspondence about this event is in the same file.

13. "Yosemite National Park Indian Arrests, Offenses and Notes," "Indian Affairs" I-3, YNPRL.

14. Beverly R. Ortiz, as told by Julia F. Parker, *It Will Live Forever: Traditional Yosemite Indian Acorn Preparation* (Berkeley: Heyday Books, 1991), 12–13. For a more general overview of Indian boarding schools, see Margaret Szasz, *Education and the American Indian: The Road to Self Determination, 1928–1973* (Albuquerque: University of New Mexico Press, 1974).

15. "Yosemite National Park Indian Arrests, Offenses and Notes."

16. Ellen Stadtmuller to W. B. Lewis, June 18, 1925, "Indians-General" I-3, YNPRL. See also later correspondence in same file.

17. C. G. Thomson to Louis Milburn, July 26, 1930, ibid.

18. T. W. Emmert to Clinton Mentzer, January 7, 1931, ibid.

19. E. P. Leavitt to Bank of Italy, Merced, October 13, 1930, ibid.

20. Sellars, *Preserving Nature,* 70.

21. Advisory Board Report for April 25, 1930, in "Briefs of Report of Advisory Board, 1928–1940," "Indian Affairs" I-5, YNPRL.

22. C. Hart Merriam, "Indian Village and Camp Sites in Yosemite Valley," *Sierra Club Bulletin* 10 (1917): 202–209; Cultural Systems Research, Inc., "Petition to the Government of the United States from the American Indian Council of Mariposa County for Acknowledgment as the Yosemite Indian Tribe" (draft), prepared by Lowell John Bean and Sylvia Brakke Vane (Menlo Park, Calif.: Cultural Systems Research, Inc., 1984), 114–115, copy in YNPRL (hereafter CSRI, "Petition to the Government").

23. CSRI, "Petition to the Government," 123.

24. W. B. Lewis, "Memorandum Regarding Indian Village," August 30, 1927, "Indians-General," I-5, YNPRL. For complaints about the living conditions in the Indian village, see CIA to L. A. Dorrington, June 10, 1927; and CIA to Dorrington, October 31, 1927; both in file 27139, Sacramento 723, "Central Classified Files, 1907–1939," Records of the Bureau of Indian Affairs, Record Group 75, National Archives, Washington, D.C. (hereafter NA RG 75, BIA CCF, 1907–39, Sacramento).

25. E. P. Leavitt to Harry Johnson, October 11, 1927, "Indians-General" I-5, YNPRL. In 1926, the National Park Service (NPS) landscape engineer criticized a house built by village resident as not being "an Indian Village type," but approved of its construction because it would eliminate other unsightly dwellings; E. P. Leavitt and Dan Hill to Stephen Mather, October, 1926, ibid. For a discussion of the rustic style in Yosemite National Park, see Robert C. Pavlik, "In Harmony with the Landscape: Yosemite's Built Environment, 1913–1940," *California History* 69 (1990), 182–195.

26. W. B. Lewis, "Memorandum Regarding Indian Village"; L. A. Dorrington to Lewis, September 1, 1927, "Indians-General" I-5, YNPRL.

27. Dorrington to CIA, September 7, 1927, NA RG 75, BIA CCF, 1907–39, Sacramento 723/27139.

28. Ibid.

29. E. P. Leavitt to Director, NPS, March 7, 1928, "Indians-General" I-5, YNPRL.

30. C. G. Thomson to Director, NPS, June 25, 1929, "Indians-General" I-5, YNPRL.

31. Thomson, "Meeting at Indian Village," July 23, 1929, "Indians-General" I-5, YNPRL.

32. Ibid.

33. Frederick G. Collet to C. G. Thompson [*sic*], August 19, 1929, "Indians-General" I-5, YNPRL.

34. Thomson to Duncan McDuffie, December 16, 1929, "Indians-General" I-5, YNPRL; Arno B. Cammerer to CIA, December 2, 1931, NA RG 75, BIA CCF, 1907–39, Sacramento 723/27139.

35. Thomson, "Special Report on the Indian Situation," to Horace Albright, January 9, 1930, "Indians-General" I-5, YNPRL; emphasis added.

36. Ibid.

37. Ibid.; Board of Expert Advisors, "Recommendations on Indian Camp," December 1929, "Indians-General" I-5, YNPRL.

38. Board of Expert Advisors, "Draft Report: Meeting of the Board of Expert Advisors, Yosemite National Park, at Yosemite Valley, April 24 and 25, [1930]," "Indian Affairs" I-3, YNPRL.

39. Thomson, "Special Report on the Indian Situation."

40. George M. Wright, Joseph S. Dixon, and Ben H. Thompson, *Fauna of the National Parks of the United States: A Preliminary Survey of Faunal Relations in National Parks,* Contributions of Wildlife Survey, Fauna Series 1 (Washington, D.C.: Government Printing Office, 1933), 4.

41. Dilsaver, ed., *America's National Park System,* 46. For an overview of these new ecological concerns and the context in which they arose, see Runte, "Joseph Grinnell and Yosemite: Rediscovering the Legacy of a California Conservationist," *California History* 69 (1990): 173–181; Runte, *Yosemite,* 161–172; Sellars, *Preserving Nature,* 91–98; R. Gerald Wright, *Wildlife Research and Management in the National Parks* (Urbana: University of Illinois Press, 1992), 14–19.

42. Wright, Dixon, and Thompson, *Fauna of the National Parks,* 4. Wright would later head up a study for the development of recreational facilities in the national parks; see Sellars, *Preserving Nature,* 139.

43. Thomson, Memorandum to the Director, July 13, 1931; quoted in Runte, *Yosemite,* 169.

44. Ibid.; Runte, *Yosemite,* 170–171.

45. Runte, *Yosemite,* 167–168.

46. For a discussion of Yosemite's fauna exhibits in the early 1930s, see Ibid., 160–168.

47. Thomson, "Special Report on the Indian Situation."

48. Thomson to Director, NPS, November 14, 1931, NA RG 75, BIA CCF, 1907–39, Sacramento 723/27139.

49. Pavlik, "In Harmony with the Landscape," 190–191; William H. Nelson to Thomson, August 2, 1935, "Indian Affairs" I-5, YNPRL.

50. Chief Ranger F. S. Townsley to Acting Superintendent Wosky, July 14, 1938, "Indian Affairs" I-5, YNPRL.

51. W. B. Lewis to Albright, January 17, 1930, NA RG 75, BIA CCF, 1907–39, Sacramento 723/27139.

52. F. S. Townsley to Thomson, February 4, 1931, "Indians-General" I-5, YNPRL; CSRI, "Petition to the Government," 137.

53. C. G. Thomson, "Memorandum to the Indians," November 24, 1931, "Indians-General" I-5, YNPRL.

54. David Parker to Mr. Collier, May 17, 1933, NA RG 75, BIA CCF, 1907–39, Sacramento 723/54915. According to park service employment records for January 1, 1932, to March 31, 1933, Parker earned just $464.75 over the fifteen-month period. This was $130

less than the average Indian wage earner over the same period of time, yet more than one-third of all Indians employed in the park earned even less. Wage records are from "Table Showing Employment Given Indians in Yosemite National Park During the Fifteen Month Period, January 1, 1932 to March 31, 1933," enclosed with a letter from Albright to Collier, June 20, 1933, NA RG 75, BIA CCF, 1907–39, Sacramento 723/54915. Comments on Parker's status in the Yosemite Indian community are based on Jay Johnson, interview with the author (telephone), Mariposa, Calif., February 11, 1998.

55. Jay Johnson, interview, February 11, 1998. I would like to thank Stella Mancillas for first bringing my attention to Bridgeport Tom's prophecy; Stella C. Mancillas, "The National Park Service and the Demise of the Indian Village, 1929–1969" (paper presented at the Western History Association Conference, St. Paul, Minn., October 1997, copy in possession of the author.

56. Albright to Collier, June 20, 1933; Collier to Parker, June 28, 1933, both in NA RG 75, BIA CCF, 1907–39, Sacramento 723/54915.

57. Quote is from Thomson to Director, NPS, July 4, 1933, NA RG 75, BIA CCF, 1907–39, Sacramento 723/54915. For reference to a similar complaint from the Yosemite Indian community, see Albright to Collier, June 20, 1933.

58. Thomson, "Special Report on the Indian Situation."

59. Thomson's remarks are reported in "Yosemite—September 21, 1932," NA RG 75, BIA CCF, 1907-39, Sacramento 723/27139. Senators Burton K. Wheeler, Elmer Thomas, and Lynn J. Frazier headed the Senate Indian Affairs Committee, which visited Yosemite as part of an extensive "Survey of Conditions of the Indians in the United States." For a brief overview of the committee's work, see Donald L. Parman, *Indians and the American West in the Twentieth Century* (Bloomington: University of Indiana Press, 1994), 86–88.

60. Harold E. Perry, "The Yosemite Indian Story: A Drama of Chief Tenaya's People," unpublished paper, 1949, file 883-07.3, "Indians-General, 1945-1954," box 970.33: I-5, YNPRL.

61. Park regulations regarding the Indian village at this time are delineated in Lawrence C. Merriam to Arno B. Cammerer, November 1, 1938, file 883-07.3, "Indians-General, 1937–1944," box 970.33: I-5, YNPRL. Jay Johnson, interview with the author (telephone), Mariposa, Calif., March 3, 1998.

62. Mancillas, "National Park Service and the Demise of the Indian Village," 5–8.

63. Jay Johnson, interview, February 11, 1998. Numbers of Indians living in the valley are cited in Mancillas, "National Park Service and the Demise of the Indian Village," 7.

64. Mancillas, "National Park Service and the Demise of the Indian Village," 9.

65. House Concurrent Resolution 108, 83d Cong., 1st sess., 67 Stat. B132 (August 1, 1953); quoted in Felix Cohen, *Felix Cohen's Handbook of Federal Indian Law,* 2d ed. (Charlottesville, Va.: Michie/Bobbs-Merrill, 1982), 171. For a broader examination of federal Indian policy in the 1950s, see Donald L. Fixico, *Termination and Relocation: Federal Indian Policy, 1945–1960* (Albuquerque: University of New Mexico Press, 1986). Development of the Yosemite Indian Village Housing Policy is discussed in Mancillas, "The National Park Service and the Demise of the Indian Village," 7–10.

66. Mancillas, "The National Park Service and the Demise of the Indian Village," 7–10.

67. Ortiz and Parker, *It Will Live Forever,* 35–36.

68. Jay Johnson, interviews of February 11 and March 3, 1998; Rebecca Solnit, *Savage Dreams: A Journey into the Hidden Wars of the American West* (San Francisco: Sierra Club Books, 1994), 290–293; Kat Anderson, "We Are Still Here," *Yosemite* 53 (1991): 1–5; CSRI, "Petition to the Government."

69. Jay Johnson, interview, March 3, 1998.

70. Under the curatorship of Craig Bates, and with the cooperation of the American Indian Council of Mariposa County, the ethnological exhibit at the Yosemite Museum now depicts native life in the 1870s—a time of great change and adaptation for the Yosemite Indians.

71. Mancillas, "National Park Service and the Demise of the Indian Village," 10.

72. Jay Johnson, interview, February 11, 1998. The text of the October 17, 1997, "Traditional Use Agreement Signed Between Yosemite National Park and American Indian Council of Mariposa County" can be found at http://www.nps.gov/yose/news_97/traduse.htm.

Notes to Conclusion

1. John Muir, *The Yosemite* (New York: Century, 1912), 256.

2. Luther Standing Bear, *Land of the Spotted Eagle* (Boston: Houghton Mifflin, 1933), xix. Iktomi Lila Sica, *America Needs Indians!* (Denver: Bradford-Robinson, 1937), 322. Iktomi is the name of a legendary Lakota trickster, and Iktomi Lila Sica is no doubt a comic nom de guerre.

3. Based on his experience as chief forester in the Bureau of Indian Affairs from 1933 to 1937, Robert Marshall proposed the creation of wilderness preserves on 4.8 million acres of Indian lands. Though sharing the paternalism of federal Indian policy in the 1930s, Marshall's ideas suggest a desire on the part of the BIA to keep some "wild" lands within reservations. Robert Marshall, "Ecology and the Indians," *Ecology* 18 (1937): 159–161; and James M. Glover, *A Wilderness Original: The Life of Bob Marshall* (Seattle: The Mountaineers, 1986), 209–211.

4. Jeffrey Mark Sanders, "Tribal and National Parks on American Indian Lands" (Ph.D. diss., University of Arizona, 1989), 32–42; Andrew H. Fisher, "The 1932 Handshake Agreement: Yakama Indian Treaty Rights and Forest Service Policy in the Pacific Northwest," *Western Historical Quarterly* 28 (1997): 187–217. The Indian New Deal produced widely mixed results throughout Indian country, and Collier's legacy remains controversial. For an overview of federal Indian policy during the New Deal, see Graham D. Taylor, *The New Deal and American Indian Tribalism: The Administration of the Indian Reorganization Act, 1934–1945* (Lincoln: University of Nebraska Press, 1980).

5. Iktomi Lila Sica, *America Needs Indians!* 315; emphasis in original. Limited to a small number of copies, the book was primarily addressed to officials in the Bureau of Indian Affairs and native leaders.

6. Ibid., 316–317, 321–379.

7. "Judge Rules Blackfeet Can Enter Glacier Free," *Hungry Horse News,* 11 January 1974, p. 6.

8. American Indian Council of Mariposa County, "Proposal to the National Park Service: The Indian Perspective on the Future of Yosemite National Park" (February 1978), 13–14, copy in the Yosemite National Park Research Library, Yosemite, Calif.; Craig D. Bates and Martha J. Lee, *Tradition and Innovation: A Basket History of the Indians of the Yosemite–Mono Lake Area* (Yosemite, Calif.: Yosemite Association, 1991), 125–126.

9. Badlands National Monument was enlarged in 1976; the entire area was designated a national park in 1978.

10. Another example of a recent cooperative effort between the National Park Service and American Indians is the joint management of Canyon de Chelly National Monument by the Navaho Tribe and the park service. Lying wholly within the Navaho reservation, Canyon de Chelly is a place of great historical importance to the tribe. The area's spectacular scenery and wealth of archaeological sites have made it important to non-Indians as well. Though Canyon de Chelly has remained largely free of Navaho sheep grazing, the

national monument was never established to preserve wilderness. Sanders, "Tribal and National Parks on American Indian Lands."

11. Stephen Hirst, *Havsuw 'Baaja: People of the Blue Green Water* (Supai, Ariz: Havasupai Tribe, 1985), 85–92, 147–148. Also see Karl Jacoby, "The Recreation of Nature: A Social and Environmental History of American Conservation, 1872–1919" (Ph.D. diss., Yale University, 1997), 356–403.

12. Hirst, *Havsuw 'Baaja,* 151–166.

13. Grand Canyon National Park Act quoted in John F. Martin, "From Judgment to Land Restoration: The Havasupai Land Claims Case," in *Irredeemable America: The Indians' Estate and Land Claims,* ed. Imre Sutton, (Albuquerque: University of New Mexico Press, 1985), 281. The Havasupai in Cataract Canyon now number over three hundred individuals.

14. For a general discussion of the Havasupai Land Claims Case, see Martin, "From Judgment to Land Restoration," 271–300.

15. The Alaska National Interest Lands Conservation Act created and enlarged thirty-five national parks, monuments, forests, preserves, wildlife refuges, conservation areas, and recreation areas. The enlarged national parks are Denali (formerly Mt. McKinley National Park), Glacier Bay (formerly Glacier Bay National Monument), and Katmai (formerly Katmai National Monument). The newly created parks were Gates of the Arctic, Kenai Fjords, Kobuk Valley, Lake Clark, and Wrangell–St. Elias. Kenai Fjords National Park does not permit any subsistence use. Likewise, Glacier Bay, Denali, and Katmai national parks do not permit subsistence use within the original boundaries of the national park or monument. For an excellent study of Alaskan national parks, see Theodore Catton, *Inhabited Wilderness: Indians, Eskimos, and National Parks in Alaska* (Albuquerque: University of New Mexico Press, 1997).

16. The history of the Alaska Native Claims Act is enormously complicated. For a concise overview, see David H. Getches, "Alternative Approaches to Land Claims: Alaska and Hawaii," in *Irredeemable America,* 301–335.

17. Catton, *Inhabited Wilderness.* Also see Roderick Nash, *Wilderness and the American Mind* 3d ed. (New Haven: Yale University Press, 1982), 272–277, 296–315; Alfred Runte, *National Parks: The American Experience,* 3d ed. (Lincoln: University of Nebraska Press, 1997), 236-258; and Craig W. Allin, *The Politics of Wilderness Preservation* (Westport, Conn.: Greenwood Press, 1982), 207–265.

18. Catton, *Inhabited Wilderness,* 212–214. At a site along the Brooks River in Katmai National Park, where grizzly bears congregate to catch salmon and tourists flock to witness this powerful symbol of American wilderness, the absence of native residents and hunters has led to a recent, almost "unnatural" explosion in the local bear population; see Ted Birkedal, "Ancient Hunters in the Alaskan Wilderness: Human Predators and Their Role and Effect on Wildlife Populations and the Implications for Resource Management," in *Partners in Stewardship: Proceedings of the 7th Conference on Research and Resource Management in Parks and on Public Lands,* ed. William E. Brown and Stephen D. Veirs Jr., (Hancock, Mich.: George Wright Society, 1992), 228–234.

19. Steve Crum, "A Tripartite State of Affairs: The Timbisha Shoshone Tribe, the National Park Service (NPS), and the Bureau of Indian Affairs (BIA), 1933–1994," *American Indian Culture and Research Journal,* 22 (1998): 117–136. Also see John G. Herron, *Death Valley: Ethnohistorical Study of the Timbisha Band of Shoshone Indians* (Denver: National Park Service, 1980).

20. Crum, "A Tripartite State of Affairs"; Timbisha Shoshone Tribe, "The Timbisha Shoshone Land Restoration Proposal" (n.d.), copy in possession of the author; Linda Greene, NPS Cultural Resource Management Specialist and Park Liaison for Death Valley National Park, interview with the author (telephone), Furnace Creek, Death Valley National Park, Calif., March 14, 1997.

21. While surreptitious native use of the Yellowstone backcountry continues to this day, some Crow men have recently contemplated a more public exercise of their treaty rights to hunt on national park lands; interviews with anonymous individuals, Wyola, Mont., July 8, 1994.

22. "Petition Submitted by the Blackfeet Tribe to Enter into a Conservation Agreement with the Secretary of the Interior to Regulate Blackfeet Rights on the Eastern Portion of Glacier National Park," January 1975; copy provided to the author by the Office of Chief Earl Old Person, Chairman of the Blackfeet Tribal Council.

23. Quoted in Brian Reeves and Sandy Peacock, "'Our Mountains Are Our Pillows': An Ethnographic Overview of Glacier National Park," draft submitted to the National Park Service, Rocky Mountain Region, Denver, May 1995, 232.

24. In compliance with the Native American Graves Protection and Repatriation Act (1990), the National Park Service is required to cooperate with federally recognized Indian tribes to identify areas and objects of cultural significance. As part of a systemwide restructuring, the park service created the American Indian Liaison Office in 1995 to facilitate the agency's relations with native peoples. While these efforts have strengthened the agency's relations with Indian tribes and fostered an appreciation for the cultural importance of national park areas, they have not addressed the legal claims that native groups might exercise in parks.

25. Quote from the 1964 Wilderness Act, Public Law 88-577, 88th Congress, S4 September 3, 1964, Section 2c; reprinted in Stewart M. Brandborg, *A Handbook on the Wilderness Act* (Washington, D.C.: Wilderness Society, n.d.), 3.

INDEX

Absaroka, 153 n.37
See also Crow
Across the Continent (Bowles), 25
Adams, Ansel, 131
Adirondack Mountains, 20, 35
Adventures of Captain Bonneville, The (Irving), 17
Agaideka, 45
See also Eastern Shoshone; Northern Shoshone
Ahwahneechee, 103
See also Yosemite Indians
Alaska National Interest Lands Conservation Act (ANILCA), 136, 137, 178 n.15
Alaska Native Claims Settlement Act (ANCSA), 136–137
Albright, Horace M.
on Blackfeet and Glacier National Park, 91, 93, 95–98, 100
and management of Yosemite National Park, 123–125
Alliance to Protect Native Rights in National Parks, 138
American Indian Council of Mariposa County (AICMC), 130, 131, 173 n.59, 177 n.70
American Indians
appeal of, for tourists, 71, 83, 86, 106–107, 112, 117, 124, 131
as defining aspect of wilderness, 4, 20, 23, 105–106
intertribal conflicts among, 16, 31, 75
perceived as threats to wilderness, 62–64, 72, 91, 94, 100, 109, 125
relocation of, from the East, 14–15
tourist fears of, in Yellowstone National Park, 57–58, 108
See also Indian wars; reservations; treaties; "vanishing Indian"; *specific native groups, treaties, and reservations*
Arapaho, 31, 49

Arapooash (Crow leader), 48
Army. *See* U.S. Army
Arthur, Chester A., 61
Asbury, C. H., 112, 113
Assiniboine, 73
Atsina (Gros Ventre), 73
Audubon, John James, 17–18

Badlands National Monument and Park, 3, 134, 135, 141 n.2
Bannock
  hunting in Jackson Hole, 65–67
  relations with Shoshone sub-groups, 46–47
  seasonal movements and territory of, 49
  treaty with United States, 50
  use of Yellowstone area (pre-park era), 42, 48–49
  in Yellowstone National Park, 51–53, 57–58, 60, 63–64, 108–109, 158 n.41
  See also Bannock Indian War; Fort Bridger Treaty; Fort Hall Reservation; Jackson Hole; Kutsundeka; Lemhi Reservation; Northern Shoshone
Bannock Indian War, 57–58
Bates, Craig, 106, 177 n.70
Beal, Lawrence, 118
Bean, Lowell, 120
beaver, 48, 51
Belden, George, 37–39
BIA. *See* Bureau of Indian Affairs
Big Trees. *See* Mariposa Big Tree Grove
bighorn sheep
  as staple of Sheep Eater diet, 47
  hunted by native peoples in Yellowstone area (pre–park era), 46
  hunted by native peoples in Yellowstone National Park, 51–53
  hunting of, by Blackfeet, 74
Billings, Frederick, 36–37
Birds of America, The (Audubon), 18
bison (buffalo)
  in Glacier National Park area, 74
  hunted by native peoples in Yellowstone area (pre–park era), 46, 152 n.24
  hunted by non–Indians in Yellowstone National Park, 52, 65
  near extinction of, 52, 60, 75
Black Elk, 3, 134, 141 n.2
Black Hills, 3, 6, 37
Blackfeet
  claims on Glacier National Park, 72, 90, 93–94, 95, 98, 99–100, 134, 139
  spiritual importance of Glacier National Park area to, 73–74, 161 nn.8–9
  in tourist promotion of Glacier National Park, 71–72, 83–86
  use of Glacier National Park (after 1910), 89, 91, 93–94
  use of Glacier National Park area (early reservation era), 75–76, 78, 80, 166 n.27
  use of Glacier National Park area (pre-reservation era), 74–75
Blackfeet Indian Reservation
  Glacier National Park area as part of, 76–80
  and off-reservation use rights, 72, 88, 93, 95
  NPS attempts to incorporate lands from, 89–90, 96–98, 100
  preservationist efforts to control hunting on, 91–92, 94
  reductions of, 75, 79–81, 95
Blackfeet Indians (et al.) v. United States, 95–96
Blackfoot (Crow leader), 31, 41, 52, 58, 154 n.55
Blackfoot Confederacy, 72
Black Hawk War, 15
Boone and Crockett Club, 64
Bowles, Samuel, 25–28, 35
Brackenridge, Henry M., 9, 11
Brott, J. H., 98
Brown, Alvis, 118, 122
Brown, Joe, 96
Brown, Lena, 107
Brown, Virgil, 119, 122
Budd, Ralph, 96
buffalo. *See* bison
Bunnell, Lafayette, 104
Bureau of Indian Affairs (BIA)
  and attempts to control off-reser-

vation rights, 64, 68, 89–90, 92–93
and efforts to reduce reservation lands, 52, 79–81
relations with NPS, 96–98, 120, 127, 134
Burke, Charles, 96

California
Indian policy of, 105, 108
management of Yosemite Valley by, 110, 112
Canyon de Chelly National Monument, 134
Carter, Jimmy, 136
Catlin, George
and first proposal for a national park, 9–10
and Indian wilderness, 11, 15, 16, 17–19, 28, 146 n.11
Chalmers, Ada, 87
Cherokee Removal, 15
Cheyenne, 31, 49
Chief Joseph, 56
Clark, Galen, 107
Cleveland, Grover, 81
Cole, Thomas, 12–13, 14, 33
Colfax, Schuyler, 25, 26
Collet, Frederick G., 122–123
Collier, John, 98, 127, 134
Comanche, 31, 45
Confidence Man, The (Melville), 30
Conger, Patrick, 60
Cooper, James Fenimore, 13–14, 15
Couer d'Alene Indians, 48
Cree, 73
Crow
relations with other native groups, 31, 56, 57
subgroups of, 47–48
and use of Yellowstone area (pre-park era), 42, 48–49
in Yellowstone National Park, 50–52, 155 n.59
See also Crow Indian Reservation; Fort Laramie Peace Conference of 1867; Fort Laramie Treaty of 1851; Fort Laramie Treaty of 1868
Crow Indian Reservation
overlaps with Yellowstone National Park, 58
reduction of, and off-reservation use rights, 52–53
Custer, George A., 108

Dakota. *See* Sioux
Davidson, Lucretia, 13
Dawes, Henry L., 149 n.55
Death Valley National Monument and Park, 137–138
deer
hunted by native peoples in Yellowstone area (pre-park era), 46, 47
hunted by native peoples in Yellowstone National Park, 52, 53
hunting of, by Blackfeet, 74, 75, 91
hunting of, by Yosemite Indians, 103, 110–111
as preferred animal in NPS wildlife management, 88
Denali National Park, 137
Devils Tower National Monument, 6
Dick, Charlie, 119
Dick, Lawrence, 118, 120
Dorrington, L. A., 120

Eastern Shoshone
relations with other native groups, 31
seasonal movements and territories of, sub-groups, 45–47, 50
use of Yellowstone area (pre-park era), 42, 48–49
in Yellowstone National Park, 51–53, 58, 60
See also Fort Bridger Treaty; Fort Laramie Treaty of 1851; Wind River Reservation
elk
Bannock hunting of, in Jackson Hole, 65–66
hunted by native peoples in Yellowstone area (pre-park era), 46, 47
hunted by native peoples in Yellowstone National Park, 51–53
hunting of, by Blackfeet, 74, 75, 91
as preferred animal in NPS wildlife management, 88, 92–93

elk (*continued*)
overpopulation of, in Glacier National Park, 99
supersedes bison as primary meat source, 49, 76
Emerson, Ralph Waldo, 11, 13, 21
Emmert, John, 99

Ferris, Thomas, 91
fire
native use of, in Yellowstone, 44, 51, 63
native use of, in Yosemite, 102
viewed as threat to wilderness, 4, 62
Forest and Stream, 62, 64
Fort Bridger Treaty (1868), 50, 67
Fort Hall Reservation
efforts to limit native movements from, 58–59, 64, 66–68
establishment of, 50, 53
Fort Laramie Peace Conference of 1867, 31
Fort Laramie Treaty of 1851, 49, 50
Fort Laramie Treaty of 1868, 32, 50
Fort Yellowstone, 57
fur trappers. *See* trappers

Galen, James, 89
Georgely, Francisco, 113
geysers. *See* thermal features
giant sequoia (*sequoia gigantea*). *See* Mariposa Big Tree Grove
Glacier Bay National Monument and Park, 137
Glacier National Park
efforts to exclude Blackfeet from, 88, 90–91, 93, 94–95
establishment of, 72, 87
and interests of Great Northern Railroad, 79, 83, 96
management of game animals in, 88, 99
NPS boundary extension proposed for, 89, 91, 96–98, 100
proposed by George Bird Grinnell, 78, 163 n.21
See also Blackfeet; Blackfeet Reservation; Great Northern Railroad; tourism; wildlife

Gold Rush (California), 102, 104
Goldwater, Barry, 136
Goodwin, George E., 91
Gordon Cumming, Constance Fletcher, 106
Grand Canyon National Park, 135–138
Grand Tetons, 44
Great Northern Railroad
and Blackfeet promotions, 71–72, 83–86
and Glacier National Park, 79, 96
Greeley, Horace, 33, 146 n.16
Grinnell, George Bird
on Blackfeet and wilderness preservation, 76, 78–79, 80–81
on game protection in Yellowstone National Park, 158 n.33
on native hunting in Yellowstone National Park, 64
as proponent of Glacier National Park, 78, 82
Grinnell, Joseph, 124

Haines, Aubrey L., 61
Harris, Moses, 63–64, 109
Havasupai, 135–136, 137
Hayden, Ferdinand V., 44, 51
Hetch Hetchy dam controversy, 115–116
Hill, Louis W., 83
See also Great Northern Railroad
Hope Leslie (Sedgewick), 13
Hornaday, William T., 91–92, 96
Howard, Maggie, 119, 122
Howell, Ed, 65
Hudson River School, 12
See also Cole, Thomas
Hutchings, James M., 106
Hyde, Anne Farrar, 25

Iktomi Lila Sica, 133–135
Indian Field Days (Yosemite), 116–120
Indian New Deal, 98, 134
Indian Territory
as cornerstone of antebellum Indian policy, 14–16
reduction and dissolution of, 29, 30
traveler's experience of Indian wilderness in, 17–18

Indian wars
  in antebellum era, 15
  as consequence of overland travel, 29, 30, 31
  as part of effort to restrict off-reservation movements, 32–33, 38, 62
  and Yellowstone National Park, 56–60
Indian wilderness
  as antebellum conception, 10–12, 22–23
  as national symbol, 13–14
  as distinct from uninhabited landscapes, 3–4, 20
Indians. *See* American Indians
Irving, Washington, 17–18, 21, 32

Jackson, Helen Hunt, 109
Jackson Hole
  efforts to exclude native hunters from, 65–66, 81
  known as Marysvale, 159 n.3
Johnson, Harry, 120, 126, 131
Johnson, Jay, 126, 127, 128, 131–132
Jones, William A., 53
Joseph. *See* Chief Joseph

Kaina (Blood Indians), 72
Kalispel (Pend d'Oreille Indians), 48, 73
Katmai National Monument and Park, 137, 178 n.18
Kay, Charles, 155 n.62
Keokuk, 16
Kootenai, 73
Kutsundeka
  relations with Shoshone sub-groups, 45
  seasonal movements and territory of, 46, 49
  See also Eastern Shoshone; Northern Shoshone

LaBreche, D. D., 90
Lacey Act, 65, 68
La Flesche, Francis, 16
Lakota. *See* Sioux
Lamar, Lucius Q. C., 61
Lamar Valley, 144 n.37
Lane, Franklin K., 86, 88, 91, 92
Lanman, Charles, 20–21
Leavitt, E. P., 119
Lemhi Reservation
  efforts to limit native movements from, 58–59, 64
  establishment of, for Shoshone, Bannock, and Sheep Eater, 50
  and off-reservation land use, 53
Lewis, Washington, 120–121
Lewis and Clark expedition, 5, 50
Lewis and Clark Forest Reserve, 81, 82
Lincoln, Abraham, 33, 102, 108
Littebrandt, William, 112, 113
Little Chief, Peter Oscar, 90, 93, 96
Little Dog (Blackfeet leader), 79, 95
Lone Wolf v. Hitchcock, 68
Ludlow, William, 61–62

Man and Nature (Marsh), 35–36
Mancillas, Stella, 129
manifest destiny
  affects of, on Indian policy, 30
  and perceptions of Indians and wilderness, 28–29
Manning, William, 65–66, 67
Mariposa (California), 131
Mariposa Big Tree Grove
  as national symbol, 35
  as part of original Yosemite Park, 109, 112
  recession of, to federal government, 110
Marsh, George Perkins, 35–36
Marshall, Robert "Bob," 177 n.3
Mary (Yosemite Indian), 107
Mather, Stephen T., 86–88, 90, 96, 116, 117
McClintock, Walter, 81
Melville, Herman, 13
Middle Park (Colorado), 26–27, 35
Miles, Nelson A., 57
Mission Indians, 103
Missouri River
  and antebellum travelers, 9, 17–18
  dams on, 10
Miwok. *See* Sierra Miwok
Mono-Paiute, 103, 106

Montana
cedes jurisdiction of Glacier National Park lands to federal government, 95
state hunting regulations and Blackfeet, 81, 92, 93
Moran, Thomas, 34
Mountain Chief (Blackfeet elder), 96
mountain goats, 74, 88
Mountain Ute, 26–27, 31, 62, 158 n.37
Mount McKinley National Park. *See* Denali National Park
Muir, John
attitudes of, toward native peoples, 23, 109
and Hetch Hetchy dam controversy, 115–116
as popularizer of outdoor recreation, 81
wilderness philosophy of, 10, 22–23, 133

Nakota (Stoney Indians), 73
Napi (Old Man), 71, 73, 75
Nation, 62
National Park Service (NPS)
dual mandate of, 116
founding of, 91, 115
and preservationist objectives, 89, 124–125, 130, 135, 137–139
tourism concerns of, 86, 116–117, 119
wildlife policies of, 88, 92, 93, 116, 124
See also Albright, Horace M.; Mather, Stephen T.; preservation; tourism; wildlife; *specific national parks and employees*
national parks
early proposals for, 9–10, 11, 16, 22, 36
See also National Park Service; preservation; *specific national parks and employees*
Native American Grave Protection and Repatriation Act, 179 n.24
Native Americans. *See* American Indians
Nettleton, Lulie, 87–88
Nez Perce
familiarity with Yellowstone, 48, 56
transmountain travel of, 46
Nez Perce War, 56, 60
Niagara Falls, 27, 35
Norris, Philetus W., 57–60
Northern Pacific Railroad
attempts to build into Yellowstone National Park, 61
interest in establishment of Yellowstone National Park, 37
See also Billings, Frederick
Northern Shoshone
and Bannock, 46
relations with other native groups, 31
seasonal movements and territories of, sub-groups, 45–47, 50
use of Yellowstone (pre-park era), 42, 48–49
in Yellowstone National Park, 51–53, 58, 60, 63–64
See also Fort Bridger Treaty; Fort Hall Reservation; Lemhi Reservation
NPS. *See* National Park Service

obsidian, 43
Oglala. *See* Sioux
Olmsted, Frederick Law, 101
Omaha Indians, 16
O'Neil, John J., 61
O'Sullivan, John, 28
outdoor recreation
gendered significance of, 87–88, 148 nn.39–40
and wilderness preservation, 35, 38–39, 78, 134

Paiute. *See* Mono-Paiute
Parker, David, 127–128, 175 n.54
Parker, Ely S., 146 n.10
Parker, Julia, 118
Parker, Ralph, 118
Pawnee, 16, 31
Permanent Indian Frontier
dissolution of, 29
as part of federal Indian policy in antebellum era, 15, 16
Piegan. *See* Blackfeet
Pikuni. *See* Blackfeet
Pinchot, Gifford, 81

Pine Ridge Reservation, 3, 134, 135
Plenty Coups, 58
Ponca, 16
predators
  eradication of, and NPS view of preferred wildlife, 88, 99, 116
  See also Grinnell, Joseph; wildlife; Wright, George
preservation, of wilderness
  as purpose of national park, 4, 61–62, 86, 88, 115–116, 124, 135, 137
  as threat to native peoples, 5, 70, 89, 93, 100, 125, 131, 133–134, 136, 138
  See also wilderness
public lands
  disposal of, 39, 149 n.51
  legal definitions of, 90
  off-reservation treaty rights on, 7, 31–33, 80–81

Race Horse, 67
railroads
  affect of, on western tourism, 25–26, 36, 102
  and impact on native peoples, 31, 37, 58
  See also Great Northern Railroad; Northern Pacific Railroad
recreation. *See* outdoor recreation
Reeves, Brian, 161 n.10
reservations
  in California, 105
  as cornerstone of federal Indian policy, 30–33
  and Indian reformers, 4, 30, 134
  and off-reservation treaty rights, 31–33, 50–51, 56–57, 66–68
  as restrictive areas, 27–28, 69
  See also Blackfeet Reservation; Crow Reservation; Fort Hall Reservation; Lemhi Reservation; Pine Ridge Reservation; treaties; Wind River Reservation
Richards, William A., 65, 67
Rinehart, Mary Roberts, 87, 164 n.8
Riner, John, 67
Rocky Mountain National Park, 145 n.3, 146 n.9
Rocky Mountains
  as distinct from eastern mountains, 21, 35
  and symbolic importance of western landscape, 26–27, 28
romanticism, 11–14
Roosevelt, Theodore, 64, 87
Rothman, Hal, 98
Runte, Alfred, 33, 125
Russell, Charles M., 79
Russell, Osborne, 18–19
  See also trappers
Rust, Jim, 126

Santee Sioux. *See* Sioux
Sargent, John Charles, 81
Sauk and Mesquakie (Fox), 15, 16
  See also Black Hawk War; Keokuk
Scoyen, Eivind T., 94, 95, 98
Sedgewick, Catharine, 13
Sellars, Richard, 119
Seminole War, 15
Senowin, Ben, 66
Se-we-a-gat, 66
Sheep Eater
  homelands of, 46–47
  relations with other Shoshone sub-groups, 47
  use and habitation of Yellowstone National Park, 50, 53, 58
  See also Lemhi Reservation; Northern Shoshone; Sheep Eater War; Wind River Reservation
Sheep Eater War, 57
Sheridan, Philip H., 62
Sherman, William T., 56
Shoshone
  regional distinctions of 157 n.20
  sub-groups of, in Rocky Mountain area, 45–47
  See also Agaideka; Bannock; Eastern Shoshone; Fort Bridger Treaty; Fort Hall Reservation; Kutsundeka; Lemhi Reservation; Northern Shoshone; Sheep Eater; Timbisha Shoshone; Washakie; Wind River Reservation

Sierra Miwok, 103, 104, 106
Sierra Nevada
  affects of American conquest on peoples and environments of, 33, 101–102, 104–105
  and symbolic importance of landscape for Americans, 35
Sigourney, Lydia, 13
Siksika (Blackfoot Indians), 72
Sioux, 3, 16, 31
Sits in the Middle of the Land. *See* Blackfoot (Crow leader)
Standing Bear, Luther, 25, 133
Stone, Forrest, 97
Swims Under, Mike, 162 n.18
Switzerland of America, The (Bowles), 26

Tabuce. *See* Howard, Maggie
Tail Feathers Coming over the Hill, 75, 93
Taylor, Bayard, 33
Telles, Lucy, 122
Tenaya, 102
Teter, Thomas B., 66, 67
Theodore Roosevelt National Park, 10
thermal features
  cultural symbolism of, for Americans, 55
  native perceptions of, in Yellowstone, 43–44, 151 n.14
  supposed native fears of, 42, 56, 58, 60
Thomson, Charles G.,
  and NPS relations with Yosemite Indians, 119, 122–124, 125, 126, 128, 130–131
  views of, on wilderness preservation, 124–125
Thoreau, Henry David, 21–22
Tichi, Cecilia, 13
Timbisha Shoshone, 137–138
Tocqueville, Alexis de, 14
Tom, Bridgeport, 127, 131–132
Tom, Leanna, 118
tourism
  in Glacier, 71, 83–86, 87
  and goals of NPS management, 88, 110–111, 116, 124–125
  as rationale for creation of national parks, 27–28, 33, 35
  in Yosemite, 117, 119, 121, 131
  in Yellowstone National Park, 53–56, 58, 60
  See also outdoor recreation; railroads
trappers
  on native peoples and western landscapes, 18–19
  and stories about Yellowstone, 42
Treaty of Medicine Lodge, 31
treaties
  and federal Indian policy, 27–28, 31, 146 n.10
  native views of, and reservations, 32–33, 49–50, 72
  See also Fort Bridger Treaty; Fort Laramie Treaty of 1851; Fort Laramie Treaty of 1868; Treaty of Medicine Lodge
Tresidder, Donald B., 117
Tukudeka. *See* Sheep Eater

Udall, Morris, 136
Udall, Stewart L., 136
U.S. Army
  as key element in federal Indian policy, 29–31, 32, 38, 56
  as manager of Yellowstone National Park, 62–64
  as manager of Yosemite National Park, 110–111

Vane, Silvia Brakke, 120
"vanishing Indian"
  and American arts and letters, 9–10, 12–13
  and expectations of tourists, 71–72, 83, 111–112
  popular myth of, and federal Indian policy, 26–27, 39, 78
  as symbol for damaged wilderness, 18, 20–21
Vest, George G., 61

Walden (Thoreau), 22
Walsh, Thomas, 93
Ward v. Race Horse, 67–68, 90, 95
Warren, Louis, 81, 165 n.16

Washakie, 32
Washburn expedition, 41–44
Waters, E. C., 69
Waterton Lakes National Park, 161 n.6
Wawona, 105
West, Preston C., 90, 95
White, Edward Douglas, 67–68
White, Richard, 30
White Calf (Blackfeet leader), 75, 80, 169 n.71
wilderness
 as defined by Indian presence, 4, 20, 23, 105–106, 134
 as escape from urban environment, 35, 87
 idealized as uninhabited place, 5, 70, 78, 98, 131, 133, 139
 identified as distinct feature of national parks, 61–62, 71, 125, 135, 138
 native peoples viewed as threat to, 63–64, 72, 109, 136
 as romantic landscape, 11–12, 36, 106
 uninhabited, as national symbol, 33–34, 71, 86, 137
 See also Indian wilderness; outdoor recreation
wildlife
 in Glacier area (pre-park era), 74–75, 76
 in Glacier National Park, 88, 89, 91, 99
 NPS management policies on, 92, 93, 116, 119, 124
 in Yellowstone area (pre-park era), 46, 47, 52, 155 n.62
 in Yellowstone National Park, 51–53, 60, 65
 in Yosemite area (pre-park era), 103, 110
 in Yosemite National Park, 109, 111, 116
 See also Lacey Act; predators; *specific animals*
Wilson, Alice James, 118
Wilson, Mary, 118
Wilson, Wesley, 119
Wind Cave National Park, 3, 134
Wind River Reservation, 50, 53, 58, 59
Wood, A. E., 109
Worcester v. Georgia, 67
Wright, George M., 124
Wyoming
 becomes a state, 67
 opposes native hunting on public lands, 65–66

Yard, Robert Sterling, 86
Yellow Wolf (Nez Perce), 56
Yellowstone (pre-park era)
 early human impact on, 43
 effect of fire on, landscape, 44
 explorations of, 41–43, 44, 150 n.4
 and ideal of uninhabited wilderness, 39, 42
 native use of, 45–50, 150 n.11
 See also Bannock; Crow; Eastern Shoshone; Nez Perce; Northern Shoshone; Sheep Eater; thermal features; wildlife
Yellowstone Lake, 42, 69
Yellowstone National Park
 bison in, 52, 65
 establishment of, 39
 Indian removal and exclusion from, 57–59, 60, 64, 68
 military management of, 62–64
 as model for wilderness preservation, 7, 61, 69–70
 monumental features of, 39, 55
 as national symbol, 7, 86, 138
 native use of, 50, 52–53, 58, 60, 63–64, 68–69
 overlaps with Crow Reservation, 58
 protection of wildlife in, 61–62, 65
 See also Bannock; Crow; Eastern Shoshone; Nez Perce; Northern Pacific Railroad; Northern Shoshone; Sheep Eater; thermal features; tourism; wildlife
Yokut, 103
Yosemite Indians
 cultural changes among, 103–104, 112–113, 170 n.15
 and NPS restrictions on residence in Yosemite Valley, 118–119, 120, 121, 125–126, 128–129, 130

Yosemite Indians (*continued*)
  and relations with tourists, 102–103, 108, 110, 172 n.34
  residence of, in Yosemite Valley (national park era), 101, 112–113, 120–124, 125–131
  residence of, in Yosemite Valley (state park era), 102, 109–110
  significance of Yosemite Valley to, 120, 127, 129, 131–132
  tourists' ideas about, 104, 105–107, 108, 109, 112
  See also Indian Field Days; Mono-Paiute; Sierra Miwok
Yosemite National Park
  establishment of, 109
  and Indian Field Days, 116–120
  Indian removal and exclusion from, 123, 127–128, 130–131
  military management of, 109, 110–111
  native use of, 109, 110–111, 112, 131
  and NPS standards for park management, 116, 119, 121
  predator control in, 116
  and preservationist concerns, 124–125, 131
  See also Hetch Hetchy dam controversy; Mariposa Big Tree Grove; tourism; wildlife; Yosemite Indians; Yosemite Valley
Yosemite Valley
  ancient human occupation of, 103
  "discovery" of, 29, 102
  early visitation to, 102–103, 105–108
  incorporated into national park, 110
  as national symbol, 27, 29, 33, 35, 86, 131
  and native management of landscape, 102
  native residence in (national park era), 101, 112–113, 120–124, 125–131
  native residence in (state park era), 102, 103–105, 110
  resources of, for native use, 103, 105, 107–108
  tourist developments in, 103, 110, 112
  See also Mariposa Big Tree Grove; tourism; Yosemite Indians; Yosemite National Park